T0251759

Architectural Conservation
Issues and Developments

Edited by

Vincent Shacklock

Managing Editor

Jill Pearce

A Special Issue of the JOURNAL OF Architectural Conservation

Routledge
Taylor & Francis Group

LONDON AND NEW YORK

First published 2006 by Donhead Publishing Ltd

Published 2015 by Routledge
2 Park Square, Milton Park, Abingdon, Oxon OX14 4RN
711 Third Avenue, New York, NY 10017, USA

Routledge is an imprint of the Taylor & Francis Group, an informa business

ISSN 1355-6207
ISBN 13: 978-1-873394-77-9 (pbk)

A CIP catalogue for this book is available from the British Library

Architectural Conservation: Issues and Developments is a special edition of the *Journal of Architectural Conservation* and is the November 2006 (Volume 12, Number 3) issue of a subscription.

Contents

The launch of the *Journal of Architectural Conservation*, April 1995, at the Centre for Conservation Studies, Leicester, attended by visitors including leading figures from Italy and throughout the UK. Foreground (r–l) Jill Pearce (publisher), Professor Kenneth Barker (Vice Chancellor), Professor Vincent Shacklock (Centre Director), Sir Bernard Feilden (Patron) and Dr David Watt (Editor).

Cover illustration Swarkestone Pavilion, Derbyshire. Built circa 1630 for the Harpur's of Swarkestone Hall, and attributed to John Smythson, the pavilion provided a grandstand view of events within the large, stone-walled enclosure (from within which this photograph has been taken). Suggestions for use of this lawned space range from jousting to bear baiting, but the most credible evidence suggests a 'bowle alley house'. Swarkestone Hall was demolished by 1750 and the pavilion was a bare shell at the time of its purchase by The Landmark Trust in 1984. Today visiting occupants gain roof access via the left-hand turret, allowing them to take in surrounding countryside, peer down upon the walled enclosure, and gain access to this holiday home's only bathroom facility in its right hand partner: a chilly walk in the wee hours of the morning, but a conversion typical of the uncompromising conservation approach taken by the Trust. Interestingly, the Swarkestone Pavilion has been and remains a popular let, and visitors remark quite positively on the roof-top jaunts. (Vincent Shacklock)

Foreword

Lessons from York Minster, 1965–1972

Sir Bernard Feilden

As this issue of the *Journal* was at the planning stages, Vincent Shacklock asked me to set out what is most likely my last contribution to a journal I helped found and launch in 1995 at his postgraduate Centre for Conservation Studies in Leicester. I was delighted that Donhead considered a refereed journal in architectural conservation a realistic proposition, and am pleased it has developed successfully with subscribers worldwide and a reputation for quality and breadth of coverage.

I was asked to write about a special project of mine and have no hesitation in choosing aspects of the 1965–1972 work at York Minster, working with Poul Beckmann – in particular, the strengthening of foundations. This gives me the chance to discuss a project that caused me the most careful thought and gave a team of rather special individuals a technical challenge of huge proportions. I believe other guest contributors to this issue of the *Journal*, drawn from various aspects of practice and education in architectural conservation, have each been asked a common question relating to conservation practice in 2006. This is a clever idea and I look forward to reading their inevitably varied responses.

York, one of the largest Gothic cathedrals north of the Alps, was first completed in 1080 with a single (16 m) wide nave. Later rebuilt in several stages, it retained much of its Norman fabric, particularly the foundations. The central tower collapsed in 1406 and was rebuilt much bigger and heavier than before, on original foundations by then already 325 years old. The wide central spans, inherited from the Normans, may have caused some anxiety to the thirteenth- and fourteenth-century builders. Whatever the reason, the nave, choir, transepts and central tower were all vaulted with timber rather than stone.

My personal, stone-by-stone, inspection of the whole Minster took place over two years and revealed a list of defects. The East End, leaned out

650 mm at eaves level. This could be restrained by shoring, but the large and widening cracks in the central tower, caused by settlement, appeared far more threatening as there was no way that the 18,000 tonnes weight of the tower could be given alternative support.

After acquiring a second opinion from the distinguished cathedral architect, Robert Potter, my fears were confirmed. Lord Scarborough, High Steward of York Minster, established an appeal to raise £2 million (an estimate I generated personally). A multi-disciplinary team was assembled with Ove Arup & Partners as structural engineers, the archaeologist supported by RCHME under the eagle eye of Sir Mortimer Wheeler, and a very able quantity surveyor.

I recall visiting the City Engineer accompanied by Poul Beckmann. We explained that we could not comply with bye-laws and building regulations. He understood our problem and waived compliance provided that I wrote him a letter, accepting full responsibility for the works! This allowed us to tackle the problem from first principles. I thought it advisable, however, to take a £2 million insurance policy, so that contributors to the appeal would get their money back if we failed. I also reflected carefully to myself that I was mortgaging my own family's future.

Shepherd Construction, a well established local contractor, was appointed on a cost-plus basis. Similarly, all consultants were paid on an hourly basis, because I considered it equitable. I was pleased that the final fee for the conservation team worked out at just over 10% of the cost of the works.

Having ascertained the soil conditions, and established that the cause of the cracking was differential settlement of the tower piers (one of which was partially founded on a Saxon cesspit), Arup's designed a way to enlarge the failing foundation laterally by placing reinforced concrete blocks in the re-entrant corners of the crossing Norman foundation walls, thus creating 14 m square footings under each pier. This was necessary, because the foundation pressures were grossly excessive and had to be drastically reduced.

The new foundation blocks needed to be linked by stainless steel post-tensioning rods, placed in holes drilled through the old foundation walls, some of them over 12 m long. Nearly 400 of these rods would be used. There were mass concrete pads underneath the reinforced concrete blocks, separated from these by Freyssinet flat-jacks which, when inflated, would pre-load the clay and thus prevent further settlement.

I called a meeting with Shepherd's men and explained the project to all of them, saying 'It is a difficult project and if you think anything is going wrong, please report it through the proper channels, and I promise you will be listened to.' The conservation team: architects, archaeologists, contractor, engineers and quantity surveyor, met every Tuesday for a site lunch of beer and sandwiches, and this promoted collaboration and teamwork.

This was just as well, because an anxious time followed. While the work was proceeding, the tower settled further, because 4 m of overburden had to be removed in order to put in place heavy concrete blocks. Despite the energy and experience of Ken Stephens, our site manager, only 8 of the 400 holes had been drilled in 6 months. I called in a mining engineer to assess the situation and he reported that we had good men, all the drilling equipment he knew about, and then some more, which he was glad himself to learn about. He felt that it was a difficult project. I decided on no more consultants' conferences as these were demoralizing Shepherd's staff. I explained to the men that we had done what we could and it was now up to the drilling teams. A night shift was set up, and the noise inside the Minster was continuous, but despite this disturbance it is worth noting that church services were not suspended all the years of our work.

The decision to proceed cleared the air. Drillers emerged with a new combination of drill bits, and from then on progress was remarkable, the holes were drilled, the rods were tensioned, and at last, the Freyssinet flat-jacks were inflated. The settlement reduced by the excavation was halted at a maximum of about 25 mm and very slightly reversed by about 1–2 mm. The central tower was stabilized. Soil tests and the calculations had shown that the pressures under the foundations had approached the ultimate limit, leaving no margin for a possible rise of the water table. The pads when loaded by inflation of the Freyssinet jacks were forced down by about 25 mm, thus preventing further settlement. The tower had only just been strong enough to survive the surgery; our hard work essential and timely.

The foundation for the Western Towers and for the East Wall were also enlarged, and the Great East window was strengthened against wind suction by a means of a system of almost invisible wire ropes, devised by Poul Beckmann of Arup's.

After seven years of work, to celebrate the successful conclusion, my partners agreed to give a dinner in York Assembly Rooms for all of those involved: drillers, mechanics, masons and other craftsmen. Anyone who had worked for one year or more was eligible. To my surprise I found that everyone had been with the team since the beginning of the rescue project seven years earlier. They could have gained more financially from building houses, but they were involved in our success. I wish I had thought of giving each a medal, so that he could tell his grandchildren what he'd done.

Years later, when I wrote to Shepherds to commiserate on the death of our project manager, Ken Stephens, they responded saying that this had been the happiest job they ever had. It seems that I was right to put my faith in Yorkshiremen. Much credit goes to Dean Alan Richardson, who gave us his trust and support, and to Lord Scarborough and Lord Halifax who led the fundraising campaign. Their efforts saved an outstanding church from crisis. I had the great pleasure to lead a team of professionals and craftsmen on a project of scale and quality seldom managed in the UK or elsewhere, to work upon one of the greatest buildings of our European heritage.

Sir Bernard Feilden
Patron of the Journal of Architectural Conservation

Figure 1 Bernard Feilden with a group of people who were largely responsible for the restoration of York Minster.

Facing up to Challenges in Architectural Conservation

Vincent Shacklock

Sir Bernard Feilden's project, described in his 'Foreword', to strengthen the foundations of York Minster in the period 1965–72, was one of the most significant and challenging interventions in a British cathedral in the last 200 years, undertaken by a matchless team, devoted to the building's protection, both professionally and emotionally. I recall visiting this project a few years after its completion in the company of my tutor Professor 'Jimmy' James, of the University of Sheffield. A York Minster guide skipped quickly over the pioneering techniques, huge technical challenges and breathtaking complexity of the work that had been undertaken, concluding it was a job well done, and noting how very pleasing it was that the Minster was back to its normal routine. Jimmy, the former government Chief Planner, described variously by Dick Crossman as 'brilliant' and 'first rate', who had chaired the high-level conference at Churchill College Cambridge first defining the need to protect historic towns, chose to wait until we were back on our journey before observing that the awe-inspiring achievement of Bernard Feilden and his team in saving this great church would, very quickly and appropriately, slide into relative obscurity. 'That is the mark of good conservation', he told me. At that time, I was not able to understand.

Tasks on the scale of the York stabilization are exceedingly rare but, across the country, in every city, town and village we daily draw upon the vision, skills and imagination of architects, engineers and craftsmen to provide timely and expert interventions in staving-off decay, rectifying failures, repairing damage, replacing features, and managing wear and tear. At Lincoln, in my role as a Fabric Council member over the last twelve years, I am very conscious of the value of education and training, having seen many of my own graduates employed on a cathedral which, less than twenty years ago, was in peril of rapid decline, but now enjoys good health, careful management, and a skilled and utterly dedicated team of craftsmen.

   Stable leadership is essential in these cases, and Lincoln's Dean and Chapter, Clerk of Works, Cathedral Architect, Chief Executive and Fabric Council members have worked tirelessly to repair, monitor and guide investment, drawing upon the best advice available, and using every opportunity to discuss and debate the programme of maintenance and individual projects. Good conservation practice frequently follows upon stable and informed management, and Lincoln's in-house building works/crafts team is highly trained, well briefed and, as a result, feels valued, secure and engaged. Individually confident that their labours contribute to the mission of the Church and are valued by clergy, worshippers and visitors alike, they enjoy the sense that they make a small but identifiable contribution to the nation's heritage with each completed task. For these reasons, they are content, hardworking, constantly refining their skills and always keen to take on challenges.

   Nationally, our objective should be the same *writ large*. The nation (through its government and institutions) must: be clear about the value it places on its architectural heritage; have systems of training, organization and representation in place; adequately fund essential work (in the interests of society and the economy as a whole); keep ahead of threats through commissioning diligent study and research; be aware of, and learn from, experience elsewhere in the UK and abroad; and make use of eager and committed groups and societies to facilitate needed work, overcome obstacles and ensure volunteer energy is channelled to achieve desirable outputs.

   Readers might observe that the papers included in this publication follow very closely this list above. Many great churches and well-loved secular buildings are managed in a manner that is mindful of this approach. Strangely, this easy lesson about benefiting from advice, support and assistance has, in very recent years, not only been ignored by the UK government, but leading politicians have increasingly questioned sector opinions, attempted to devise their own philosophies, and proposed policies based on prejudice and suspicion of the sector's attitudes and ambitions.

   This publication appears at just the moment we expect a White Paper on Heritage, a document whose lengthy gestation has been marked by greater uncertainty, confusion and political ambivalence than we have witnessed for some years. In the lead-up to the White Paper, some government politicians, seemingly uncomfortable with physical heritage as defined in recent decades, have wrestled with ideas intended to re-define the nation's heritage in a manner that would be more inclusive of society's many social and ethnic groups. Bob Kindred's very fine paper, 'What Direction for Conservation? Some Questions', enquires into the origins of this suspicion and its potential implications, commenting on it with a skill and deftness of touch borne of many years careful observation of aspiring politicians and their methods. His assessment is elegant and well-reasoned: significant figures in government believe heritage problems have been largely resolved; the impending scale of investment required for the 2012 London Olympics demands fundamental reconsideration of expenditure in the Department of Culture, Media and Sport (DCMS); and to achieve this, transfer of funds away from historic assets is appropriate and essential.

With a policy review imminent and legislation likely to follow, the paper from Professor Malcolm Airs, 'Protecting the Historic Environment: The Legacy of W. G. Hoskins', assesses the development of the English landscape over several centuries; and, in particular, Hoskins' stimulation of debate at a time when few peers in this field existed. The final chapter of his groundbreaking book *The Making of the English Landscape* was devoted to the changes and threats he perceived, and this plainly caused him great distress. I read his work as a student in 1971, felt his despairing, pessimistic gloom and took his concerns at face value. But times and society were already changing, and these would bring a raft of new interests, controls and protections into place over the following twenty years that would steer countryside, town and village landscapes more positively. Malcolm Airs, a one-time research student of Hoskins, handles this with the skill and judgement we would expect from such a distinguished academic and commentator.

Adam Wilkinson's paper, 'SAVE Britain's Heritage and the Amenity Societies', points out the importance of having well-organized and committed amenity societies, capable of turning their minds and resources to issues swiftly, bringing understanding and expertise, and ensuring that issues are not overlooked in statutory or other decision-making. Although many countries have voluntary organizations representing public interests in heritage protection, the United Kingdom benefits from a broad and unique range of period-based and theme-based bodies, which are more often than not able to work collaboratively where circumstance requires. The UK is particularly unusual in the manner in which certain key societies have been drawn into the statutory planning process as formal consultees. Effective conservation in the UK has depended in large measure upon a committed, educated and inspired voluntary sector, capable of handling casework on a daily basis, turning resources to urgent threats and undertaking or commissioning research projects as circumstances demand. SAVE Britain's Heritage is part of this movement, yet distinct in its methods. It is long established but abhors convention, undertaking a particular role of its own in a bold fashion now recognized as the hallmark of its operations.

Washington DC-based Donovan Rypkema, a prominent development consultant dealing with the re-use of historic structures, compares and contrasts UK and US models for architectural conservation, observing that the aims are the same but the methods decidedly different. His paper, 'The American Contrast', is timely and thought-provoking, vividly depicting, and accounting for, a successful working model in the US which is, in marked contrast to our own, largely local, bottom-up, incentive-driven, and private sector. Contrasts in approach could hardly be greater, but returning UK visitors often remark on the stunning success of individual private and community-based conservation projects, and Rypkema improves our understanding of how this occurs. Rypkema is also an author and professor at the University of Pennsylvania, and his balanced and informed assessment strengthens this publication's significance as a commentary on UK practice.

We are fortunate to have two outstanding scholars of historic gardens and landscapes, David Lambert and Jonathan Lovie, the former and current conservation officers of the Garden History Society, pool their skills to provide a paper examining achievements in defining, identifying and protecting gardens and landscapes in recent years under the title 'All Rosy in the Garden? The Protection of Historic Parks and Gardens'. The authors, however, express important, perhaps grave concerns in some areas, as English Heritage cuts resources, the Heritage Lottery Fund tightens its belt, and the National Trust finds itself struggling to sustain maintenance income. Their fear is that parks and gardens will suffer disproportionately in this dilemma. At a local level, the authors advise us that planning authorities remain hamstrung by a dearth of conservation expertise and are struggling to cope with known threats and proposals, while not having the time needed to identify and understand their own stock of historic parks and gardens. If this is not sufficient to worry us, Lambert and Lovie warn that large-scale and harmful proposals for altering historic parks and gardens are, once more, on the rise.

But times do change, and we deal with the world as it is. Today we do have a professional Institute for Historic Building Conservation and a reasonable spread of professional training courses relating to architectural conservation and heritage management, but John Preston's paper, 'The Context for Skills, Education and Training', reveals a frightening absence of craft skills, no holistic analysis of skill needs for the sector, and no convincing case made for building conservation, repairs and maintenance as considered separately from general construction. Efforts to improve skill levels through accreditation have, as yet, failed to make any significant impact. We are bereft of comfort or basic reassurance that the development industry can devise a strategy, let alone a successful plan of implementation, for the development and sustenance of craft skills.

Professor Peter Brimblecombe and Dr Carlota Grossi of the School of Environmental Sciences, University of East Anglia, look at the research-base of practice in architectural conservation, calling for a more comprehensive research agenda that balances past achievements with emerging issues of the future. Their paper, 'Scientific Research into Architectural Conservation', reveals weaknesses and worrying gaps in research on conserving the built environment along with proposals on what future strategies must encompass. Issues in relation to research dissemination are of particular concern.

An Appendix, 'The Listing of Buildings' by Bob Kindred, provides a general outline of the way in which listed buildings in England are currently protected.

I remember first seeing the opening pages of John Delafons' *Politics and Preservation: Policy History of the Built Heritage, 1882–1996* when it was first published, and commenting to a colleague that I had found it impossible to focus upon the opening paragraphs of this outstanding review without being drawn repeatedly, to the impressively well-selected cartoon on the left-side page.[1] A burly Viking warlord, in battledress, stands at the head of a column of marauders – their axes, swords and spears a potent declaration of destructive intent. A longboat on the beach disgorges additional, heavily armed marauders. But observing a public notice board, the leader has come to a stop and, resting his weapon on the floor as he digests the message, explodes in angry frustration: 'Bugger me... This is a conservation zone!' We are, of course, encouraged to imagine the Viking horde taking reluctant heed of the notice, returning to their oars and departing to pillage and destroy an unfortunate settlement elsewhere, one unencumbered by such a restrictive statutory designation.

It was with this cartoon in mind that during an informal gathering between IHBC members last year, I asked innocently whether preventing unauthorized harm within conservation areas was more difficult now than in the past. Had historic buildings been safer then, simply by virtue of their appearance on the Statutory List? Did inclusion in a Conservation Area in those days make owners more cautious than now of making even minor changes without careful check with the local planning authority? Conclusions were difficult to draw, but the feeling tended to the view that conservation area designation and/or listing was once rather more of a check on hasty, unplanned and unauthorized alterations. It was felt that twenty years ago, the development process was more predictable. A generally slower pace of change, planning officers' frequently impressive local knowledge, the more positive regard in which they, conservation specialists, in-house architectural advisors, and other 'experts' were held, had meant that a measure of discussion and guidance usually informed the owner's understanding, and led to a less harmful building intervention.

Since then, our society has changed enormously. The plastic window phenomenon, together with its stable-mates – plastic eves, guttering and doors – and a bewildering stream of television programmes encouraging them to 'd.i.y. their way' to greater comfort, convenience and resale value, has force-fed the idea that property is an investment, to be coaxed to ever greater return. So, the gathering hesitatingly concluded, rogue owners of historic properties had been slightly fewer in number, or at least easier to identify, influence and manage. But though this applied at the local and domestic level, there was less sign of it applying more widely; furthermore, and rather worryingly, it was suggested that the time available to a conservation or planning officer to give advice to a building owner was, in many cases, less with each passing year.

In the ten years since Delafons skilfully drew our attention to the achievements and failings in our protection of our architectural heritage, his publication remains the most impressive critique of strategy and policy in the UK. But the nature of our economy and society has changed more in the last ten years than in the previous thirty. Government in the UK, at all levels, now functions in constant flux, and architectural conservation finds itself within a turbulent confusion of organizational structures, policies and practices. Managing English Heritage is unquestionably more complex in this climate.

It was Michael Heseltine, then Secretary of State for the Environment, responding to a proposal from Maurice Mendoza, Director of the of the then Ancient Monuments and Historic Building Directorate, who set in place the generally helpful process of change leading to the creation of English Heritage. A resultant 1982 document – *Organisation of Ancient Monuments and Historic Buildings in England: The Way Forward* – set out ideas and provided reassurance for many readers.[2] The new body would manage around 400 monuments, including castles, abbeys, burial mounds, hill forts, and so on, presently in the care of the Secretary of State. All held with the brief to bring them more 'alive' so that they were better visited, appreciated and understood. It would also: make grants for the preservation of historic buildings, areas and monuments; act as adviser to the Secretary of State on listing and scheduling; guide matters of policy; and inform statutory decision-making. Heseltine had the advantage of being, more than most other ministers of the period, visually literate, understanding of architectural heritage, and interested in the environment as a whole.

The argument that professional expertise and greater commercial-mindedness would be more evident if undertaken by an agency, rather than a government department, was not lost in the radical Thatcher years. Peter Rumble, the first English Heritage chief executive, confirms that the parliamentary debate did not really cover the rationale for division of responsibilities, and seemed to revolve around the assumption that if an action benefited an individual or organization, it could be given to an agency, but if, on the other hand, it could be deemed harmful, particularly in relation to property rights, it should rest with a minister accountable to

Parliament.[3] The National Heritage Act was given Royal Assent in May 1983, with the agency we would soon know as English Heritage coming into place on 1 April 1984. By the same day in 1986, staff committing themselves to continued employment in this new organization ceased to be civil servants, and, with the demise of that thorn in the Prime Minister's side – the Greater London Council (GLC) – the GLC's Historic Buildings Division became, albeit reluctantly, part of the new body.

English Heritage's mission has always fallen across several ministries and so has been obliged to engage, inform, educate or defer to a range of ministers. This can never have been an easy task, particularly when so many secretaries of state and ministers have come into the job, particularly in recent years, without much interest in the field. Michael Heseltine and Peter Brooke seemed to have had an intellectual understanding of the UK's physical heritage. Dick Crossman was interested, well informed, and had a keen eye for a building of quality. Despairing at the lack of a good private secretary, he wrote in his diary 'Thank God...' on news that he acquired the services of John Delafons.[4] But even Delafons might have had difficulty with the cross- and multi-ministry practices of the modern government and civil service.

I have some sympathy for the English Heritage director who recently and positively expressed the view that

> ...given the cross-cutting nature of the historic environment, which has never been easy to keep within administrative boundaries, it was... particularly encouraging when in 2003 ODPM [the Office of the Deputy Prime Minister] and DEFRA [the Department for Environment Food and Rural Affairs] joined DCMS as joint signatories of English Heritage's funding agreement ... It was probably this that marked the historic environment's real coming of age as a proper concern for government.'[5]

I risk being considered dreadfully naïve, but I cannot find real evidence of conservation being a concern for government in a manner that might attract critical praise; as for the identification, protection and support of the nation's architectural heritage being dependant upon effective collaboration between three Whitehall ministries, collaboration even within one is sometimes quite a cause for celebration.

For this special publication, more than a dozen figures who have made their contributions as politicians, policy-makers, commentators, writers and so on have gamely agreed to respond to the question: 'It has been said that anxiety pervades the British conservation sector in 2006. What changes would you most welcome or most strongly resist, and why?' There are some intriguing replies, and I am grateful to them all for taking up this challenge. It is interesting to see the VAT issue come up again, and to see reinforcement of some of the principal themes emerging from the main papers.

As for what more we might have done in this publication, it would have been good to include greater coverage of issues in relation to churches, world heritage, contemporary design intervention in historic buildings, and matters of new-build architectural design in important historic areas. The Editorial Advisory Board does, however, plan to return to these areas in due course.

Finally, I should record the warm thanks of the publisher and all members of the Editorial Advisory Board for the advice, guidance and encouragement provided by Sir Bernard Feilden over the last dozen years, covering the *Journal*'s early development, editorial advisory board appointments, and its establishment here and abroad as a publication of choice in the broad field of architectural conservation. Though he has now stepped down from his former active role, we are all most pleased he has agreed to remain our Patron.

Biography

Professor Vincent Shacklock MA, DipLandArch, IHBC, FRTPI, FRSA
Vincent Shacklock is Dean of Architecture, Art and Design at the University of Lincoln. He played a key part in establishing the Journal and is a founding Editorial Advisory Board member. Following work in three local authorities, he ran a private practice before being appointed Director of the multidisciplinary Centre for Conservation Studies at DMU, Leicester. He has led projects on the conservation of various historic buildings and gardens in Italy, and has been a member of Lincoln Cathedral's Fabric Council during the last twelve years of extensive repairs. He has lectured in the US and Italy on historic architecture and gardens.

Notes

1   Delafons, J., *Politics and Preservation – A Policy History of the Built Heritage 1882–1996*, E & FN Spon, London (1997).
2   Department of the Environment (Organisation Development Division), *Organisation of Ancient Monuments and Historic Buildings in England: The Way Forward*, HMSO, London (1982). This Division was also responsible for co-ordinating the setting up of English Heritage and the abolition of the Ancient Monuments Board and the Historic Buildings Council for England in 1984.
3   Rumble, P., 'The Creation and Early Days of English Heritage', in *Conservation Bulletin: English Heritage – the First 21 Years*, Issue 49, Summer 2005.
4   Crossman, R. H. S., *The Crossman Diaries. Selections from the Diaries of a Cabinet Minister 1964–70* (edited by Anthony Howard), Hamish Hamilton and Jonathan Cape, London (1979), p. 80.
5   West. J., 'England's Heritage. The Changing Role of Government', in *Conservation Bulletin: English Heritage – the First 21 Years*, Issue 49, Summer 2005.

Personal Perspectives

It has been said that anxiety pervades the British conservation sector in 2006. What changes would you most welcome or most strongly resist and why?

I am concerned at some of the ways we seem to measure our success as modern custodians of historic fabric; if it is by the eloquent rhetoric of conservation or the amount of media coverage, we are on dangerous ground; if by the amount of reorganizing, rationalizing and streamlining (however necessary and beneficial some of this may be) we will be sidetracked by the activity. Only by looking often and systematically with informed eyes at the condition of our historic buildings, monuments and landscapes will we ever know the truth.

We must also look carefully at ourselves. If there are historic property managers who barely know the significance and importance of their site, if there are architects and surveyors who do not understand the difference between the repair and maintenance of new buildings and ancient ones, if there are contractors who are not able to match the scope and quality of work of their predecessors in past centuries, if there are archaeologists who are more familiar with the reality of their computer screens than the fabric of their sites and if we let our advancing technology seduce us into believing that recording, investigating and researching relieve us of our duty of care, then we should all be concerned.

Professor John Ashurst D.Arch RIBA
Ingram Consultancy, Salisbury

One welcome new area of recent conservation thinking is the deeper understanding of historic towns and cities and how to care for them better. Building conservation has made great strides forward in the last generation but care of our urban fabric has been very poor. At last we are realizing how rich culturally and socially are our towns and how we can undo some of the damage inflicted on them by earlier insensitive interventions, 'Characterization Studies' have just begun to be a useful tool in which the

history, quality and value of towns is first understood before designers and transport planners are let loose. The process is only in its infancy but has great potential to realize the enormous cultural inheritance of our urban places.
Alan Baxter BSc, FIStructE, MICE, MCONSE, Hon.FRIBA
Alan Baxter & Associates, Consulting Engineers, London

Ever extended 'listing', general public enthusiasm and greater understanding of architectural characteristics have helped improve the conservation of vernacular buildings over the past 30–40 years. At the same time there is a danger that over-intrusive regulation and over-detailed control by conservation officers could work against a generally sympathetic public attitude. More and better education in this field, as in others, is the key to successful conservation of vernacular buildings in the future.
Dr Ronald W. Brunskill OBE, MA, PhD, FSA, Hon.D.Art
Author on vernacular architecture, former Commissioner of English Heritage

There are achievements to prompt optimism; trends which deserve anger. The latter includes the mindless weakening of professionalism in local planning authorities and other bodies (influenced by 'managers', various charlatans and political ineptitude). Added to this is the complacency of professional bodies. Conservation should long have been accepted as an integral part of planning. We must resist threats to professionalism, we should promote it.

Changes that would be welcomed: a national and effective grant system for the repair of historic buildings; and an acknowledgement that good contemporary design can be integrated within our historic buildings and townscapes.
John Dean Hon.D.Art
Fellow and past President of The Royal Town Planning Institute

We need much more tangible and long term Government commitment to our built heritage which is increasingly overshadowed by a 'quick win' culture in a department that seems obsessed by sport and gambling. We need an understanding by an anxious architectural profession that accreditation is not an establishment plot to secure restrictive practice but an essential path to protect our most valuable inheritance through a higher level of understanding. In an theory driven architectural education system, history and conservation are the Cinderella subjects which need proper recognition, alongside a greater understanding that the marriage of ancient and modern is the greatest challenge in the art of architecture but, when

successful, can produce the most brilliant results. I want to see national recognition of the huge contribution that good conservation and creative re-use makes to our civilization – and that conservation is as much to do with sustainability and skylines as it is to do with mortar mixes!
George Ferguson
RIBA Past President; Chairman of Acanthus Ferguson Mann; Director at Academy of Urbanism

I would like to see a strengthened understanding among decision-makers at all levels of how vital a good physical environment is in shaping our mental and spiritual well-being – both for individuals and society – and how much informed conservation has to contribute to that. This has special relevance to church buildings, which embody and support the memories of their communities as well as their present aspirations in a very tangible way.
Paula Griffiths MA (Oxon)
Head of the Cathedral and Church Buildings Division of the Archbishops' Council

I had the rare privilege of serving in the Department of the Environment on three separate occasions, twice as Secretary of State. I would like another chance to put right things I got wrong. I would transform the listed building and heritage arrangements. There would be two categories of licence.
> *Category A*: Individual owners in either the public or private sector would be able to look after their buildings on estates within guidelines that would be drawn to high conservation standards.
> *Category B*: Professional bodies or firms would be licenced to both approve and supervise such work.
Local authorities would have to receive notice and detail of work proposed and would have defined powers to intervene. Abuse by either of the two categories of licence holders would lead to revocation. Provided Category A licence holders reached agreement with a Category B licence holder and served appropriate notice on the local authority in a timely manner they would be free to manage their property as they saw fit. No one would prejudice their licence with the risk that they would revert to the previous arrangements which are expensive, slow moving and conformist.
  I wish I had thought of it at the time.
The Rt Hon the Lord Heseltine CH

We must not be obsessed with identifying the heritage; there are now around 500,000 listed buildings in England, which is enough, at the present rate of demolition, to last for at least 5,000 years. Conservation areas have been degraded through over use – almost everywhere of any interest has been designated, along with much that is in no way 'special'. Public opinion is at present generally favourable to conservation; but pendulums swing and views will change. The real challenge is going to be to manage that process of change.

Good conservation is thus simply a facet of good planning – every proposal needs to be assessed in the light of its likely effect on all the relevant land and buildings, historic and otherwise. So there is no need for special consent regimes, and special conservation duties embedded in legislation, as these give a special place to historic buildings that is not justified, but also – paradoxically – marginalize the heritage. The whole system has undoubtedly become far too complicated.

The most welcome changes would, therefore, be for the conservation sector to become more open to allowing, and indeed encouraging, the enhancement, rather than just the preservation, of historic buildings and areas; and for the planning system as a whole (of which conservation is just a part) to be radically simplified.

Charles Mynors FRTPI, MRICS, IHBC
Barrister, author, diocesan chancellor, visiting professor Oxford Brookes University

It's not all doom and gloom! Public interest in history and architectural heritage is exceptionally high – look at this year's Heritage Open Days and the *History Matters – pass it on* campaign. What I would like to see change is the official response to such enthusiasm. I'd like to see it recognized as bringing real and tangible benefits to society, not just nice for those who can. I'd like to see this played through to policy and public spending decisions – a longer tunnel at Stonehenge and more conservation staff in every local authority, for example. The quality of our built environment is a huge economic draw and provides enormous quality of life benefits – too often we are still brushing these considerations aside instead of focusing on the real, tangible value that they bring.

Fiona Reynolds
Director General, The National Trust

From the fund raising point of view, which is what I do and have been doing for Hackney for years, I am not conscious that the year 2006 has been a lot more dreadful than any of the last ten years. Abolishing VAT on repairs is still a proper, worthwhile and hopeless ambition. It would be the one measure which would show political commitment to renovation, recycling and repair rather than waste and neglect.

Eating up fields and allowing old buildings to fall into dereliction is encouraged by current legislation. God knows, there are 'initiatives', but they seem to be to let back gardens designated brown field sites and acres of useable building bulldozed or rebuilt with shoddy replacements. Having worked on *Restoration Village* this year I think we should find more money for semi-private developments generally and smaller perhaps less distinguished vernacular buildings all round, but then, perhaps I would think that, wouldn't I?

But I believe that the huge amounts of cash poured into that iconic Guano extravaganza, Tyntesfield, for example, might be better spent repairing the whole county of Somerset's barns or preserving the façades of Weston Super Mare. For the same reason I long for some sort of institution to actively promote new thinking about re-use. Small conservation groups are often left struggling for ideas to save a building. We need to match need with available space. Someone needs to be coming up with good commercial ideas for old buildings and passing them around, especially places of assembly. But much of this is politically difficult. Lottery money cannot apparently be used to prop up private houses. Which brings us back to VAT. It is tax that incentivizes the majority. We need a break. It seems so obvious I suppose we must squat in the rubble and despair while ministers blame the EC.

Griff Rhys Jones
Presenter of the television programme 'Restoration'

I do not know who said anxiety pervades the British conservation sector but that is certainly not how I would characterize the mood of voluntary organizations concerned with our heritage. The voluntary conservation movement, much admired in other countries, is probably stronger now than it has ever been. The degree of co-operation, the enthusiasm and capacity to take on new projects, the willingness and determination to find ways of engaging with a much wider audience are greater than ever I can remember. Where there is cause for concern it is that the Government does not adequately appreciate what the voluntary sector is achieving, nor the extent to which the public in general cares about their heritage.

John Sell CBE

I would be thrilled to see the Government enthusiastically accept virtually all the recommendations in the Culture, Media and Sport Select Committee Report on Heritage, and acknowledge that, in cutting expenditure on heritage, they have overlooked the benefits in terms of regeneration and place-making, tourism, education, civic pride, social stability and sustainability that flow from the care of historic buildings and landscapes.

I hope the Secretary of State will soon discover the delights of visiting historic places and notice the large numbers of family groups and foreign visitors enjoying themselves whilst learning how our rich and complex history has fostered the New Labour virtues of equality, democracy, creativity and tolerance.

Les Sparks OBE, Dip Arch, DipTP, RIBA, MRTPI, FRSA, HonD.Des
English Heritage Commissioner and former CABE Commissioner

Conservation thinking is now mainstream and accepted as part of the central focus of managing and moderating change. If anxiety pervades the conservation community it may perhaps be due to a perception that heritage is now owned by society rather than the preserve of the specialist. Television has broadened awareness and the internet has made technical information available to all.

A sensitivity to our past, but recognizing the need for appropriate change (including sustainability) should become the cornerstone of the education and ethics of every built environment professional. I would hope that future built environment education, at both an undergraduate and specialist level, would include a thorough understanding and appreciation of our past as a contributor to a better future.

John Worthington
Founder DEGW, Graham Willis Professor University of Sheffield

Education is the key to understanding, cherishing, enjoying and conserving historic places. I'd particularly like to see a real commitment from national and local government to primary school 'citizenship' projects on the history on our doorsteps, and how we can help those old places to lives new lives. For a tiny investment a whole community's eyes can be opened by their own children. Conservation is a public benefit, but it will not be a high priority for democratic government unless the wider public really values it, and tells our politicians how much we care.

And what scares me? It is ignorance – people who can't see or won't see the time dimension in where they are. They may be cost-led politicians or managers, under-trained construction people, or building owners consigning their history to the skip because they don't understand or value it.

John Yates
Chair, Institute of Historic Building Conservation (IHBC)

Protecting the Historic Environment

The Legacy of W. G. Hoskins

Malcolm Airs

Abstract

Landscape history as an academic discipline effectively began with the publication in 1955 of The Making of the English Landscape *written by W. G. Hoskins. It was a highly influential book, shaping the way that a whole generation has managed the inherited historic environment. With 50 years having passed since the publication of the book and major legislative changes on the horizon, this is an appropriate moment to reflect on the development of the conservation movement, and to speculate on some of the challenges that need to be faced over the next half-century. The evolution of the mechanisms for statutory protection under the impetus of changing public opinion and academic scholarship are explored. New political imperatives such as regeneration, sustainability and social inclusion are discussed, alongside the impact of the media and the National Lottery in creating a climate where positive change is embraced as part of a process of greater understanding of the environment.*

The English landscape

W. G. Hoskins (1908–1992) had a deeply rooted sense of place. He was born in Exeter, and he returned to live there while he was still teaching at Oxford and then at Leicester. His outrage at the post-war redevelopment of his beloved city prompted him to become an early activist in the cause of conservation. He was a founder member of the Exeter Civic Society, and in his capacity as its Chairman, he was co-opted onto the City Council. However, his outspoken allegations of malpractice drew a writ from almost the whole of the planning committee, and he was forced to settle out of court. The experience left him sadly disillusioned about contemporary civic responsibility and ended his public campaign, although he continued to express his views in private. Hoskins failed in his campaign to save significant parts of historic Exeter – partly because there were no effective systems in place to protect historic buildings and areas 50 years ago, and partly because society in general had not yet come to recognize the necessity of managing the past in a way that would benefit future generations. The final chapter of his groundbreaking book on *The Making of the English Landscape* was devoted to the landscape of the 1950s, and its state caused Hoskins to take a deeply pessimistic view. 'Since [the late nineteenth century] and especially since the year 1914', he wrote, 'every single change in the English landscape has either uglified it or destroyed its meaning, or both'.[1] He despaired at the depopulation of the countryside and the disappearance of the traditional village way of life, and railed at the destructive effects of modern farming and the decline of the market town. He complained about the ravishment of the landscape by the extractive industries and its defilement by new airfields and arterial roads. Above all, he mourned the passing of the country house and the values that it represented.

The evolution of control

This lament for a disappearing world was probably justified, at least on Hoskins' terms. In its place, however, has come a remarkable transformation in how we view our responsibility for the past – a transformation that is now so deeply embedded in our consciousness as to give a fair degree of optimism for the future. Thus from the perspective of 2006, it comes as something of a shock to reflect that the controls for the protection of the historic environment were so minimal in 1955. The notion that it was everyone's right to do what they wished with their own property was still largely unchallenged, despite the campaigning of Ruskin, Morris and others in the previous century. True, the state had begun to make modest inroads on that right as far back as 1882 with Sir John Lubbock's Ancient Monuments Act, but it covered only uninhabited ruins, and simply gave the state the opportunity to take them into guardianship with the owner's consent. A meagre total of 68 structures were included on the original list, but even so the bill was attacked on its passage through the Commons as an assault on private property 'to gratify the antiquarian tastes of the few at the public expense'.[2] As far as inhabited buildings were concerned, the approach that was chosen was not to protect them, but to simply record and classify them in the prevailing tradition of scientific enquiry. In 1908, the Royal Commission on Ancient and Historical Monuments and Constructions of England (subsequently shortened to the

Royal Commission on the Historical Monuments of England) was charged with making an inventory on a county-by-county basis, but such was the definition of 'historical' that only buildings erected before 1714 could be included. It was as late as 1963 that this arbitrary terminal date was lifted. Thus, the volume for the City of Oxford published in 1939, for example, gives only a perfunctory notice to Gibbs' Radcliffe Camera (1738) and Wyatt's Oriel Library (1788), and ignores altogether Cockerell's masterpiece of the Ashmolean Museum (1841).

The Commission was absorbed into English Heritage in 1999, its incomplete inventory long since abandoned. Although the county volumes that it had published did contain very short lists of monuments that were considered worthy of protection, it was only the comprehensive post-war planning system introduced by the Town and Country Planning Act of 1947 that began to seriously address the protection of inhabited buildings. Under the Act, the relevant minister was charged with the responsibility of compiling a list of buildings of special architectural or historic interest, and the listing system as we now know it was born. An advisory committee was set up, of which Hoskins became a member. Fourteen investigators were appointed to compile the list, which was divided into Grade I and II buildings, and a supplementary list of Grade III buildings with no statutory force. At some unspecified date, a further category of Grade II* buildings was introduced, which might have satisfied the fastidious standards of the advisory committee, but which has only served to confuse the general public ever since.

Listing was a slow process. The number of investigators was soon reduced, and by 1951 only 12,496 buildings had been listed. This had risen to 85,753 by 1964, but full coverage of the whole country was not completed until two years later.[3] The list was firmly elitist in the types of buildings that were included, with the majority of the vernacular buildings that had been the subject of Hoskins' essay on the Great Rebuilding[4] put into the supplementary lists, with most of the buildings on the statutory list dating from before 1840 (Figure 1). Nevertheless, according to the architectural standards prevailing at the time, it was a notable achievement. For the first time, the state had taken stock of its built heritage and given public notice of its special status. Once the process had begun, it soon became clear that listing was not a finite exercise. By 1968, developing perceptions of what was of historic value, driven by the relentless urban redevelopment that Hoskins had sought to prevent in Exeter, led to pressure to expand the list to include under-represented categories. Some slow progress was made, but by 1980 only four investigators were employed on the task.

Then there was a dramatic change. Shocked into action by the publicity surrounding the demolition of the Firestone factory, Michael Heseltine, as

Figure 1 Brookhampton Farmhouse, Oxfordshire. A typical vernacular farmhouse, reflecting regional architectural traditions. Listed Grade II in March 1975, demolished April 1978 following a public inquiry. Replaced by a development of 12 houses.

Secretary of State for the Environment, initiated an accelerated resurvey of the whole country, with much of the fieldwork being carried out by private contractors rather than civil servants. The result was a considerable expansion in the number of buildings deemed worthy of inclusion in the list, so that today there are approximately 500,000 entries, and even buildings as young as ten years old are eligible for listing as long as they are of outstanding quality and under threat. On the whole, individual Secretaries of State with responsibility for the list have never flinched from the statutory duty to include any building that they have been advised is of special architectural and historic interest. The sole exception is in the politically charged area of public housing. Successive ministers have failed to grasp this particular nettle, and there can be little doubt that this major phenomenon of twentieth-century life will be inadequately represented in anything like its original form in 2055. It is already too late in the wake of the individuality that came with the 'right to buy' in the 1980s.

Initially, the controls that accompanied the list were pretty ineffective. The owner of a listed building simply had to give six months' notice to the local planning authority of their intention to demolish or significantly alter the building. If the authority wished to resist the proposal, they had to issue a Building Preservation Order (BPO), which could be appealed to the Secretary of State via a public inquiry. It was a cumbersome system that put the onus entirely on the local authority. Not surprisingly, very few authorities had the expertise or the will to resist damaging change or demolition, and only 187 BPOs had been confirmed in the first eleven years of the 1947 Act. By 1964 that number had risen to 344, but even so that was a paltry response to the vast amount of destruction that was taking place in those years.[5] One of the few planning authorities that took its responsibilities seriously was the London County Council, which had an honourable tradition of caring for historic buildings going back to the London Survey Committee of 1898. Its expansion as the Greater London Council (GLC) in 1966, taking in the whole of Middlesex and parts of Surrey, Kent and Essex, led to the creation of a Historic Buildings Division under the leadership of W. A. Eden, who had been a fellow member of the listing advisory committee with Hoskins. It had a staff of around 100, including a number of historians whose principal task was to research the detailed history of those buildings which were the subject of public inquiry under the BPO procedure. Although no other county could match the expertise of the GLC, a few of them, such as Essex and Hampshire, set up similar but smaller departments to operate the system. The rest of the country, however, saw no need to appoint specialized conservation professionals.

By 1967, the inadequacies of a system which theoretically protected individual buildings of acknowledged value, but allowed their surroundings to be devastated, was causing increasing concern to a body of opinion led by Duncan Sandys in his role as Chairman of the Civic Trust (Figure 2). The legislation had then been in place for 20 years and something like 120,000 buildings had been listed, but still the development industry was changing the appearance of the landscape at a pace and in a way that many regretted. The concept of the 'familiar and cherished local scene', which might not contain many buildings of listable quality but which nevertheless had a distinctive character meriting preservation and enhancement, was promoted and enshrined in the Civic Amenities Act of that year. Local authorities were required to identify such areas within their boundaries, and to designate them as conservation areas. Apart from controls over trees and advertisements, the Act did not give them any additional powers at first, but even so, it was of great significance because it devolved the responsibility for identifying whole areas worthy of protection from central government to local politicians (Figure 3). This assumed greater importance once control over the demolition of all buildings in a conservation area was introduced in the following year. As the number of locally designated conservation areas has steadily increased to the current 10,000 or so, one suspects that Ministers have regretted the decision ever since. Certainly they did not follow the same path when the designation system was expanded to embrace historic parks and gardens in the 1980s, and historic battlefields in the 1990s. The compilation of the registers of these distinct elements of the historic landscape was firmly lodged with the experts of the government quango, English Heritage, even though the designation itself offers no statutory protection.

Figure 2 The Wellington Inn, Manchester. Under the direction of F. W. B. Charles, the last surviving timber-framed building in the city centre was raised by nearly 5 feet to accommodate the underground car park for the Arndale Centre in 1976–7. The Centre itself was demolished in 1996 following the IRA bombing, and the Wellington has once again been redeveloped to give a more sympathetic setting.

Figure 3 High Street, Dorchester-on-Thames. One of the first conservation areas in Oxfordshire to be designated following the 1967 Civic Amenities Act.

Control of demolition in conservation areas was introduced as part of a radical change in the protection regime by the Town and Country Planning Act 1968. The principal element of that Act was the introduction of a system of listed building consent. For the first time, the owners of listed buildings needed specific permission for any alteration or demolition that they wished to carry out. Instead of the onus being on the planning authority to actively prevent undesirable works, it was now firmly on the applicant to justify their proposals, in a climate where there was an officially expressed 'presumption in favour of preservation'. Moreover, uniquely in the planning system, it became a criminal offence to carry out works to a listed building without consent, with imprisonment as an option in the most serious cases. At last the legislation had real teeth and, despite detailed modifications in subsequent acts, this remains the system under which we currently operate.

Public perception

On the whole it is a system which commands general respect, and a recognition that its provisions act for the common good of society as a necessary restriction on sectional interests. This seems to be the reasonable conclusion to draw from a MORI poll conducted in 2000, where 98% of those polled agreed with the proposition that the historic environment is a vital educational asset which enables people to understand their history and their own identity. A meagre 23% felt that we preserve too much, which must be seen as a remarkable endorsement of the protection system. As many as 88% thought that the historic environment was important in creating jobs and boosting the economy.[6] This opinion was reinforced by a recent study on *Tourism and the Heritage Dividend* reported in November 2004. Tourism in England is largely driven by our history. It is worth £76 billion a year to the economy, and is set to grow to £100 billion by 2010. It is the fifth-largest employer in the UK; in Greenwich alone, to take one example, the lure of this World Heritage Site accounts for 25% of all jobs – employing 6,000 people and contributing £327 million to the local economy.[7] In the light of these statistics, it is not surprising that 85% of those polled thought that the historic environment has a key part to play in promoting regeneration in our towns and cities. Undoubtedly, it all depends on how the questions are phrased, but nevertheless, the MORI poll represents a major shift in public opinion from the belief prevalent earlier that conservation is a minority interest that inhibits economic growth.

The change is so significant that it is worth considering some of the landmarks that helped to bring it about. The demolition of two very different Victorian buildings in 1962 outraged informed opinion and galvanized the media into berating the government for its failure to intervene. J. B. Bunning's Coal Exchange in the City of London was demolished for a road-widening scheme that was never achieved, and Philip Hardwick's magnificent Euston Arch was needlessly sacrificed to the redevelopment of Euston Station. There can be little doubt that these two pointless demolitions helped to stimulate a growing demand for more effective controls, which was to be met only six years later in the 1968 Act. In the same way, the demolition of the Firestone building on the outskirts of west London over a Bank Holiday weekend in 1980 was not only the catalyst for Heseltine's accelerated resurvey, but it also awakened an interest in the architecture of the twentieth century, which in turn led to an appreciation of its vulnerability (Figure 4). A key role in this widening of attitudes towards the architectural legacy of the past has been played by the national amenity societies. It is possible to chart the stages in the development of public attitudes to our historic legacy by the dates when these societies were founded. The Georgian Group was formed in 1937, in the wake of the demolition of Robert Adam's Adelphi in the previous year. The Victorian Society came into being in 1958, and its membership was considerably enhanced by the row over the Euston Arch. SAVE Britain's Heritage was founded in 1975 as a media lobbying group following a shocking exhibition on the destruction of the country house held at the Victoria and Albert Museum the previous year, and the Twentieth Century Society emerged from the rubble of the Firestone factory.

Figure 4 The Hoover factory, Perivale, 1931–5 by Wallis, Gilbert & Partners. Listed following the demolition of the Firestone factory designed by the same architectural practice. The factory closed in 1982, and was given a new lease of life in the 1990s by conversion into a supermarket.

The effectiveness of their lobbying would not have been possible without a parallel development in academic scholarship in the field of architectural history. The great, knighted triumvirate of Summerson, Pevsner and Colvin might not have seen themselves as stalwarts of the conservation movement, but a younger generation of historians soon began to forge the inevitable links between this new scholarship and the need to preserve the physical evidence on which it was based. Architectural history, as an academic discipline, has traditionally concentrated on what has been labelled 'polite' architecture – buildings designed by professionals and commissioned by the elite – but more recently there has been a similar increase in knowledge about the buildings of ordinary people. The Manchester school under the inspired direction of Ron Brunskill gave the latter subject academic respectability, but much of the scholarship in this field has been driven by the recording activities of dedicated amateurs, who have made significant contributions to knowledge. The vernacular of farmhouses and cottages and urban tenements that Hoskins described so eloquently in his above mentioned paper on the Great Rebuilding is now recognized as the major factor in the regional distinctiveness that characterizes the historic landscape.

At a more popular level, the immense power of the visual media, particularly television and the cinema, has helped to stimulate an extraordinary interest in buildings of all periods. It is not just the success of such obvious programmes as *Time Team* and *Restoration*, but the whole plethora of makeover and property-themed series, and the popularity of costume dramas where historic buildings form the essential backdrop, that are responsible. A new awareness of our surroundings is reflected both in public debate and in private conversation. This debate has been fuelled in the last few decades by the perceived failures of the architectural profession itself – a perception that is only now beginning to dissipate. The disastrous performance of many modern buildings, and the arrogant refusal of many architects to acknowledge any existing sense of place or context in new development, has alienated public opinion. However, there are encouraging signs that this general cynicism is beginning to change, as a batch of landmark buildings by architects with an international reputation has captured the imagination of a wide public (Figure 5). It is clear that the 'Bilbao factor' has become a potent force in the regeneration of whole areas that only a decade ago were considered to be in terminal decline. Iconic buildings have brought fresh prosperity to some parts of the country where traditional industries have disappeared. As the success of Manchester and Tyneside and the Eden Project in remaking the post-millennial landscape becomes ever more apparent, this trend is bound to increase – although whether the market can sustain an infinite number of landmark developments is a different question that still remains to be

Figure 5 Lloyd's of London insurance market and offices, 1985, Richard Rogers and Partners. An early example of one of the iconic buildings which has helped to restore public confidence in the architectural profession.

answered. However, it can be predicted with confidence that many of these buildings will be listed and cherished by 2055, as part of England's new architectural heritage.

What is equally heartening is that the regeneration of many cities has been achieved as much by the creative reuse of the existing building stock as by the building of new architectural masterpieces. This is partly due to the global realization that the resources of the world are finite, and that future generations face an energy crisis of major proportions unless steps are taken to address it now. Sustainability is a word that trips off the lips of every Western politician at every available opportunity nowadays, and it is at the very heart of sensitive conservation. The energy embedded in existing building stock is capital that has been already spent. Before it is discarded, careful thought needs to be given to the possibility that it can be utilized for a new purpose that gives it added value. Adaptive reuse, rather than wholesale redevelopment, has the potential to retain the identity that makes places special whilst still meeting the demands of the modern world. Tate Modern is a striking example of such a development, with the monumental grandeur of Gilbert Scott's power station being as much part of its success as the modern interventions of Herzog and de Meuron. The same is true of the Baltic Gallery in Gateshead, and of the warehouses in Manchester that have brought young people back to live in the city. Conservationists have been promoting the value of new uses for old buildings for a generation or more. Now the rest of the world has caught up.

Funding

One of the major drivers of this phenomenon has been the National Lottery. In the first ten years of the Heritage Lottery Fund, the amount of money available for the repair and enhancement of the country's heritage has been totally transformed. It has spent £3 billion on nearly 15,000 different projects. Approximately a third of that sum has gone on rescuing buildings at risk that would have stood no chance of survival without substantial injections of capital. The rest has gone on a concerted effort to improve the quality of the environment which deliberately targets areas of social deprivation over a remarkably broad front.[8] A fundamental objective is that every applicant has to demonstrate how their scheme will ensure that everyone can learn about, have access to, and enjoy their heritage. This sense that the historic environment belongs to all, and not just the educated middle classes, is a challenge that will change radically the ways that it is managed in the future. In a multicultural society, everyone has a different perception of what is valuable from the past. Many groups in that society feel powerless and excluded. They do not feel that their historical contribution is being properly celebrated. This is an issue which will need to be addressed directly if the England of 2056 is to reflect truly the diversity of contemporary society. All the signs suggest that this more holistic approach is slowly being recognized.

Managing change

An understandable weakness in Hoskins' analysis in his final chapter was his implicit assumption that changes in the landscape of the past had all been somehow beneficial, whereas the contemporary changes that he saw happening all around him were universally evil. In today's world, this is an unacceptable position to adopt. Hoskins, more than anyone else, demonstrated that the landscape is characterized and enriched by centuries of change and modification. Logically, if the results of past changes are to be celebrated, then the inevitability of future change must also be accepted. It is quite simply impossible to fossilize the landscape. Indeed, it is highly undesirable to attempt to do so. What it does mean is that change has to be managed with great care if future generations are to continue to enjoy those elements of the landscape that currently give pleasure. Perceptions of what is significant are constantly being revised, and will continue to change in the future. Those revisions are largely based on knowledge. The better something is understood, the more highly it is valued. Knowledge grows exponentially, and it is impossible to know how it will develop over the next 50 years, but it can be confidently predicted that the tools available for managing the landscape in 50 years' time will be based on a greater understanding of the richness and complexity of the landscape.

The most important tool is, of course, the system for protecting the landscape which has been laboriously established over the last half-century. Its worth is now generally accepted, and it is likely that it will survive for the next 50 years, albeit in a modified form. Some of those modifications are already in train. Instead of a confusing variety of listings, schedules and designations, each with a different regime of control, there will be a single register of historic sites and buildings with a unified system of consent. Additions will continue to be made to the register, but the process will become more transparent by consultation with owners and third parties, and a right of appeal against inclusion on the register will be established. Provided that society is prepared to devote the resources necessary to achieving it, the management of the system will become increasingly professional and responsive to public concerns. Fifty years ago, the concept of a 'conservation professional' did not exist. Now, the Institute of Field Archaeologists (IFA) and the Institute of Historic Building Conservation (IHBC) are both playing a recognized role in the maintenance and promotion of professional standards. In the fullness of time, it is likely that they will amalgamate into a single, more powerful body in response to the notion that the historic environment is indivisible above and below ground.

The members of the combined Institute will be faced with a bewildering variety of pressures for change and development, such as: the continuing shift in the centres of population; the seemingly insatiable demand for new housing in open countryside; the decline in traditional methods of farming; the search for new sources of energy supply; and changes in transportation, to name just a few. All of them are likely to affect radically the appearance of the traditional countryside, just as the turnpike roads and the canals did in the eighteenth century, followed by the railways in the nineteenth century, and the motorway system in the twentieth century. The Great Rebuilding, when it comes, will be a regional phenomenon, just like Hoskins' rebuilding in the sixteenth and seventeenth centuries. Unlike his thesis, this time it will be fully documented in government White Papers and Planning Policy Guidance notes, public inquiry statements and polemical articles in the quality newspapers. Futhermore, if the ill-conceived Pathfinder Programme reaches fruition, it will be accompanied

Figure 6 Taynton, Oxfordshire. A traditional threshing barn converted into a house, but with the loss of its defining characteristics and its farmyard setting.

by a Great Demolition of unwanted houses, largely in the deprived areas of the industrial north.

Without the exercise of careful stewardship, whole groups of building types will disappear or be so totally altered as to be rendered incomprehensible. Traditional farm buildings, especially threshing barns, are particularly vulnerable. Some of them undoubtedly will find benign new uses that respect the logic of their original form, but many more will be changed beyond all recognition, losing any semblance of their historical purpose (Figure 6). It is a process that is already well underway in some parts of the country. The whole landscape of industrial buildings, on which the wealth of so many cities was founded, is another vulnerable category. Conventionally, this is seen as a challenge for those areas where the heavy manufacturing and extractive industries were centred, but in reality it is a problem that also affects regional industries like malting or brewing, which once sustained small market towns in agricultural areas. This can be illustrated by the example of Wallingford, a small historic market town on the Thames, which less than 40 years ago was able to support a brewery, several maltings, two agricultural foundries, a commercial laundry, a boatbuilding yard and a furniture factory. Now there is no industry at all, simply houses and service providers of various sorts.

There are other classes of buildings whose loss is quite simply unthinkable, and it is impossible to believe that future generations will not ferociously fight to keep them substantially in their present form. Parish churches are an obvious example, but strangely enough in an egalitarian society, country houses are another. Hoskins thought that they were doomed, and so they seemed in the 1960s and 1970s. Large numbers have indeed been lost, and many others have been mutilated by institutional use, which at one time seemed to be the only alternative to demolition. However, over the last decade or so the finances of the hereditary aristocracy have rallied on the back of rising land values and a booming art market to such an extent that ancestral homes are being lovingly restored, and country estates have once again become a symbol of financial success for entrepreneurs and businessmen. Historic houses have come out of institutional use and are reverting to single-family ownership, and a surprising number of new country houses are being built, albeit in a traditional architectural idiom (Figure 7).

One of the major challenges facing the management of the existing building stock will be to find sufficient skilled labour to maintain and repair it in traditional ways. The heritage building sector is calculated to be worth £2 billion a year, but it is suffering from an acute shortage of building skills.[9] Although it would be foolish to be complacent, this shortage is not felt so much in the specialized crafts, where it is now easier than 40 years ago to find thatchers and conservation carpenters, lime plasterers and even specialists in regional materials like cob and dry stone walling. Many of them are educated people who have developed a love for traditional methods, and have therefore created small businesses to ensure their survival. These specialized building companies will continue to expand as long as conservation professionals create a market demand. It is in the wider building trade that the shortage of skills is most marked. With the demise of the apprenticeship system, knowledge of the ordinary trades of bricklaying, carpentry and roofing is not being passed on. Various initiatives are being formulated to address the problem but, in a society that is seeking to place 50% of its young people in university education, it is likely to remain a predictable challenge for the next generation.

Many other totally unpredictable challenges will undoubtedly arise, but it is possible to be reasonably confident that future generations will meet them with a greater understanding of what the historic environment is and what it means to people. That is not a bad legacy from the last 50 years, and much of the credit is due to the scholarly foundations laid down by Hoskins. Despite the political gulf that separated them, he would have agreed with the sentiment expressed by Chairman Mao in his *Little Red Book*: 'the nation that forgets its past, forfeits its future'. It is a lesson that we must continue to heed.

Figure 7 Henbury Hall, Cheshire, 1985, by Julian Bicknell for Sebastian de Ferranti. It replaced a Victorian house demolished in 1957. (John Ashdown)

Biography

Malcolm Airs MA, DPhil, FSA, FRHS, IHBC
Malcolm began his career as a historian with the GLC (1966–74) before working as a conservation officer with South Oxfordshire District Council (1974–91). He has been a Commissioner of the Royal Commission on the Historical Monuments of England and has served on the advisory committees of English Heritage, the Heritage Lottery Fund, the Conference on Training in Architectural Conservation (COTAC), the Architects Registration Council of the United Kingdom (ARCUK), and the National Trust. He is a past President of the Institute of Historic Building Conservation and is currently Professor of Conservation and the Historic Environment at the University of Oxford.

Notes

1 Hoskins, W. G., *The Making of the English Landscape*, Hodder & Stoughton, London (1955), p. 25.
2 Delafons, J., *Politics and Preservation – A policy history of the built heritage 1882–1996*, Spon, London (1996), p. 25.
3 Ibid., p. 79.
4 Hoskins, W. G., 'The Rebuilding of Rural England: 1570–1640', *Past and Present*, No. 4 (November 1953).
5 Delafons, *op.cit.*, p. 79.
6 A summary of the findings is contained in *Power of Place*, English Heritage, London (2000).
7 *Sunday Times*, 21 November 2004.
8 Heritage Lottery Fund, *It's Your Heritage* (2004).
9 Countryside Agency, *English Rural Crafts, Today and Tomorrow* (2004).

The Context for Skills, Education and Training

John Preston

Abstract

Conservation skills remain in short supply within the construction industry. Progress in increasing the number of professionals with conservation expertise has generally been disappointing. The challenges relating to availability of conservation craft skills have been quantified, but as yet there is no holistic analysis of skills needs for the sector. Potential support for skills development has not materialized because the government has not been convinced of the case for building conservation, repairs and maintenance to be considered as a sector in its own right. Efforts to improve skills levels through accreditation have, as yet, failed to make significant impact.

Introduction

Key ingredients for good architectural conservation include: conservation-aware building owners; builders competent in traditional construction; and professionals able (according to the circumstances) to specify appropriate repairs, and/or to sensitively integrate new work within historic settings. The latter may involve anything from a minor upgrading of services in existing buildings (Figure 1), to the creation of new buildings of scale, proportion and detail complementing and enhancing an historic ensemble.

The UK is far from having this ideal combination of ingredients. Up to 50% of UK building and construction work comprises alterations, repairs and maintenance, but construction training is heavily focused on new builds. Training in conservation tends to be more of an optional extra, undertaken only by those with a specific interest, despite the need for such training having been clear for many years.

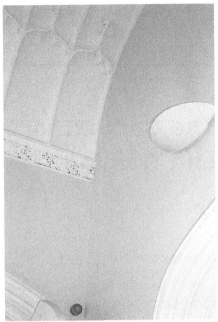

Figure 1 The need for awareness. New heating system and lighting in Grade I listed building, Kings College, Cambridge. The right illustration is as inserted by client and engineers without reference to architect. The left illustration is after remedial works designed and supervized by the architect.[1]

Slow progress

In the first issue of this journal, Sir Bernard Feilden recalled a 1975 report recommending that every UK architectural office should have at least one person qualified in architectural conservation, and that specialized courses should be set up to provide them.[2] Thirty years on, far less progress has been made than we might have expected or hoped. The number of specialist conservation courses remains small in relation to the need, and conservation barely features in architecture degree programmes. Feilden returned to the issue in 1999, listing sixteen different types of professional involved in conservation in a table and relating them to the International Council on Monuments and Sites (ICOMOS) Training Guidelines.[3] He identified five professions as being involved in all of the ICOMOS-defined tasks: architects, conservation officers, conservators, landscape architects and surveyors. In 2002, Orbasli and Whitbourn noted the lack of conservation training in architecture courses, the customary specialization in conservation at postgraduate level, and the higher demand for this in times of reduced economic activity.[4] They raised particular concerns in relation to the architecture curriculum's heavy focus on design-based skills, and the lack of a framework for controlling standards of conservation training and the conduct of conservation professionals. In 2004, authors from Historic Scotland and English Heritage outlined developments in conservation accreditation, their own organizations' requirements for lead professionals on grant-aided projects to be accredited, and the work of the pan-professional 'Edinburgh Group'.[5] The Historic Scotland and English Heritage requirements for employing accredited professionals are responses to 'fundamental difficulties... experienced in seeking to achieve appropriate quality and standards in a number of Historic Building Repair Grant Scheme cases'.[6]

An unrecognized demand

Projects reached by Historic Scotland and English Heritage repair grants constitute only a tiny proportion of the total work carried out every year. England alone has nearly 500,000 listed buildings, and over 9,000 conservation areas containing around 1 million dwellings. This is a significant proportion (about 6%) of the total building stock. Nearly 5 million buildings in England (over 21% of the total) are pre-1919. England has over 14,000 listed places of worship, with Anglican churches alone making up 45% of Grade I listed buildings. In 2004–5, over 38,000 applications for change (nearly 35,000 applications for listed building consent and 3,400 for conservation area consent) were determined, and over 25% of the 645,000 planning applications made had conservation implications. Together, this suggests an annual total of around 200,000 projects affecting historic buildings and/or areas, but even this does not provide a full picture. It does not include schemes which require building regulation approval only, repairs requiring no formal approval, and works to churches exempt from secular control.

Conservation and the construction industry

Conservation has traditionally been seen as a very small specialist part of the building and construction industry, and considered separately from general repairs, maintenance and alterations. The Construction Industry Training Board (now CITB ConstructionSkills), dominated by large contractors, provides little voice for the small builders who carry out most works to traditional and historic buildings. This picture has improved since 2004 through the creation, by English Heritage and CITB ConstructionSkills, of the National Heritage Training Group (NHTG). The NHTG's 2005 *Skills Needs Analysis for the Built Heritage Sector In England* is a key report showing the economic significance of the sector.[7] Total annual spend on listed buildings was estimated to be £1.72 billion for the 12 months before the survey, rising to £1.85 billion for the 12 months after. For pre-1919 buildings, the estimates were £3.54 billion rising to £3.68 billion. The report highlighted: shortages of skilled sub-contractors (with 6,590 craftspeople needed over 12 months to meet immediate skills shortages); a lack of workers and trainers in the 30–45 age group (with a consequent risk of far greater skills shortages as older and more experienced workers retire); and a lack of apprentices due to the government's focus on maximizing the number of school-leavers entering university degree programmes, rather than undertaking skills-based training.

Parallel issues for repairs and maintenance had been highlighted in work for Maintain our Heritage, which proposes an English counterpart to the Monumentenwacht in Holland. This work has identified problems such as the lack of skills for understanding the significance of heritage properties, and inadequate practical understanding of the use and performance of traditional materials.[8]

The NHTG report did not consider conservation professionals, so missing a vital opportunity to provide a sector-wide picture of skills needs and availability. We do know (from English Heritage) that 1,700 out of 5,400 Royal Institute of British Architects (RIBA) registered UK architectural practices profess some conservation expertise, but there is no clear and independent basis for establishing this. Ten years after conservation accreditation schemes were set up in England and Scotland, there are still less than 300 RIBA Architects Accredited in Building Conservation (AABC), and fewer than 70 Royal Institute of Chartered Surveyors (RICS) members accredited in building conservation. The Society for the Protection of Ancient Buildings (SPAB) gave evidence to the Culture, Media and Sport Select Committee which painted a bleak picture:[9]

> Far too few professionals working on historic buildings have had any specialist conservation training. Much of the Society's casework is prompted by the actions of professionals with little or no grasp of con-servation ideas or practice. Building conservation at even its most ele-mentary level forms no part of undergraduate courses in architecture, surveying etc. The large debts facing newly graduated architects means that few can afford further specialist training. Many do not believe they need it…

The need for conservation awareness and training is now greater than ever, not just among the traditional professions, but also across a broader professional landscape. The scope of Feilden's analysis could now be widened to include structural engineers (identified as a priority by English Heritage), regeneration professionals, urban designers, building control surveyors, mortgage surveyors, facilities managers, and many others whose professional practice affects historic buildings or their settings.

Figure 2 The need for skills in local materials. A good materials match, let down by excessive pointing in works to listed flint estate wall, Rousdon, Devon.

The vernacular challenge

A vital part of the British landscape is the range of regional and local vernacular building materials and traditions, arising from the country's varied geology.[10] Preservation and enhancement of this regional and local character depends just as much on appropriate repairs to lesser vernacular structures (cottages, outbuildings, boundary walls) and on appropriate construction of new buildings in historic contexts, as on the repairs of buildings defined by statutory measures. These local materials and construction techniques, which give historic places their character, were initially used because they were easiest and cheapest. Now it is these very local inputs that require special materials and skills.

The National Heritage Training Group was set up specifically to improve the situation, but has not yet achieved changes on the scale needed. Just one Centre of Vocational Excellence (the Building Crafts College) specializes in conservation. There is no regional network of centres focusing on local building and conservation skills. In England, the government has set up Regional Centres of Excellence in Regeneration, as part of its Sustainable Communities initiative. These Centres of Excellence (as with Regional Development Agencies and Sustainable Communities as a whole) have focused almost exclusively on the creation of new communities, not the sustaining of existing communities and their historic environments.

The government context

Historic environment conservation sits across a series of governmental fault-lines. One fault-line is between 'construction' and 'culture': education, information and archive management, all of which are essential to support conservation, come under 'culture' rather than the 'construction' industry.[11] Another, this one within 'culture', is between new creativity and archaeological interpretation of the past, as distinct categories. The historic environment, its creative challenges, and the skills and resources needed to manage it fall through the gap in between the two, almost unnoticed in governmental and other strategies for arts and culture. The Department for Culture, Media and Sport (DCMS) has conspicuously failed to realize the significance of these fault-lines, let alone try to bridge them. Within the sector, its Heritage Protection Review is considering the need for closer links between conservation officers and archaeologists, and 'new skills and greater capacity' within local authorities.[12] This bridging of gaps will be welcome if it happens, but (on its own) will have minimal impact on the wider need to build bridges across major governmental fault-lines, to change perceptions, and to develop momentum for improving skills across the sector as a whole.

Government skills initiatives have been focused on creating a more competitive workforce by means of 25 employer-led Sector Skills Councils, which are expected to develop occupational standards and training programmes in agreements with their workforces.[13] CITB ConstructionSkills has been confirmed as the Sector Skills Council for Construction: this development has reinforced the domination of large contractors. The focus on new work has been further increased by separating repairs, maintenance and property management into the remit of another Sector Skills Council: Asset Skills. Cultural heritage aspects of conservation, including artefact conservation and archaeology, are the responsibility of yet another Sector Skills Council: Creative and Cultural Skills. Just when the NHTG report appeared to offer unequivocal evidence of an economically significant sector warranting special consideration, this fragmentation between different Sector Skills Councils has made it harder than ever to promote the conservation, repairs and maintenance of traditional buildings.[14]

This challenge has been made harder still by the government's encouragement of up to 50% of school-leavers to attend university. It rejected the Tomlinson Report's recommendations for enhanced vocational qualifications, which might have boosted entry into conservation trades. There has been little support for mature students and others interested in moving to a new career in conservation work. Postgraduate conservation courses have developed, but remain pitifully few in relation to the potential need. There is a wide and developing range of undergraduate and postgraduate courses in broader conservation subjects, but these vary greatly in content and the extent to which they could provide a grounding for young practitioners.[15] At craft and trade level, efforts to develop Mastercraft qualifications in conservation have made painfully slow progress; there have been problems in getting both suitably trained teachers and potential trainees from busy employers.

The conservation sector

A fundamental concern is that there is no UK-wide lead body for the heritage field, and no clear voice for tackling UK-wide issues. English Heritage led the formation of the National Heritage Training Group (NHTG), but missed the opportunity for a holistic approach covering the full conservation sector spectrum from trades to professionals. This was hard to understand at the time, and it seems even more inexplicable now. It would have been so much better if the NHTG's 2005 report had covered the sector as a whole. This failure is, sadly, coupled with strategic failures to see, and seize, potential opportunities. Did English Heritage and Historic Scotland, as the sector lead bodies, promote the economic and workforce productivity importance of conservation repairs and maintenance, and hence the need for special consideration within the developing Sector Skills context? The outcome suggests not. How much better might it have been if English Heritage, in particular, had engaged with the process, achieved representation within the Sector Skills Development Agency (the overseeing body), and promoted a cross-sector approach.

Another missed opportunity, on the part of English Heritage and DCMS, has been the failure to secure a significant profile for the historic environment within the major government skills initiatives (the Egan Review and the Academy for Sustainable Communities) for delivering the review of the planning system and the Sustainable Communities agenda. English Heritage has tried to raise awareness of conservation among local politicians and public sector professionals through HELM (Historic Environment Local Management), but its combination of a website and one-off training events does not as yet seem to have had any significant impact on the target groups. By contrast, Historic Scotland (HS) has given a very positive lead through the technical conservation research and education work of the Scottish Conservation Bureau. It was Historic Scotland's quality-audit of grant-aided work that led to its requirement for conservation accreditation of lead professionals on its projects, and it has been HS which has brought the conservation professions together in the Edinburgh Group to explore accreditation. Some of the most significant work in building up a UK-wide picture has been carried out not by the national statutory bodies, but by the Heritage Lottery Fund (HLF). Its 2000 report highlighted the key issues; now the HLF has put its principles into practice by investing £7 million in training bursaries.[16] The most complete mapping of skills for the sector so far has been provided by Heritage Link, a consortium of voluntary bodies.[17]

The professions

What of the professions? Feilden criticized the failure of RIBA to appreciate the creative element in architectural conservation. In recent years, two of which were with a conservation architect as its president, the RIBA has entered into partnership with the AABC Conservation Accreditation scheme. Will this welcome change of approach continue? There is a long way to go yet, judging by the RIBA's comment to the Culture Media and Sport Committee that:[18]

> Architects are highly-trained and have a special set of skills that they can use to benefit the historic environment. Conservation architects work all over the country to restore historic properties and extend their viability...

This seemed complacent at best, and was in sharp contrast to the SPAB view quoted earlier.[19] The RIBA's comment might have had more validity if the Institute was making efforts to promote conservation within mainstream architecture courses.

The RIAS (Royal Incorporation of Architects in Scotland) has been more committed, with its long-standing accreditation scheme, at three levels, which both recognizes the importance of conservation and provides progression for architects gaining experience. The Royal Incorporation of Chartered Surveyors introduced its conservation accreditation to put its members on a par with architects in terms of access to conservation grants. The Royal Town Planning Institute (RTPI) has produced a conservation good practice guide, but for non-specialists. Within these institutes, there have been conservation interest groups; however, these have not achieved major influence, either within their own institutes or in their external collective activities (such as the Construction Industry Council). The RICS and RTPI did not even respond to the Culture Media and Sport Select Committee (see Notes 9, 18, and 20).

The Institute of Historic Building Conservation (founded in 1997) has 1,500 members and is tiny by comparison with the RIBA, RICS and RTPI. It is the only professional body providing a clear voice for architectural conservation.[20] Its journal, *Context*, has given extensive coverage to developments and issues relating to conservation training and accreditation.[21] The IHBC has 'affiliate' and full membership categories, which provide an entry route for interested professionals. The IHBC's membership is multidisciplinary, including planners (33%) and architects (23%); around 40% of its members have specialist qualifications in building conservation.

Accreditation

Formal training is only part of the picture; successful conservation depends on a combination of academic training, practical experience and insight. Accreditation in conservation requires the submission of project evidence to show that these requirements have been achieved. The Edinburgh Group's Framework for Accreditation has the potential to provide a pan-professional basis for assessing competence in architectural conservation. The Group is developing web-based 'refresher' units to help professionals prepare for accreditation; these will have to be combined with practical experience. The Group's framework (unlike the RIAS's accreditation and the IHBC's membership categories) does not provide an entry route and ladder for interested professionals. Instead, Historic Scotland and English Heritage have focused on the perceived need of clients for a single common standard of professional, whatever the discipline. This single high level of accreditation does not cater for 'general practice' professionals who do conservation work; it has to be questioned in the context of the urgent need to draw more professionals into conservation. This need will become ever greater and more urgent as current shortages of conservation professionals are magnified by a 'retirement time bomb' (evidenced by analysis of the IHBC, which has a large proportion of members aged 50 or over) similar to that for the trades highlighted by the NHTG report.

The scope, development and relevance of accreditation have also been heavily constrained by a focus on building repairs, rather than the broad spectrum of conservation work. This has been driven partly by professions devising schemes (in the case of the RICS and AABC/RIBA schemes) to meet English Heritage and Historic Scotland's requirements for grant aid, and partly by English Heritage and Historic Scotland's reluctance to intrude on the RIBA's territory in terms of design. This focus purely on

Figure 3 Refurbishment involves both repairs and design. Refurbishment at Willow House, Cambridge (listed Grade II*) included new external insulation, with all windows other than the one under the balcony replaced with double-glazed units and moved out to new positions.[22]

repair (and within that on the limited number of projects grant-aided by English Heritage and Historic Scotland) has had very limited impact, by comparison with the need, on improving the availability of conservation skills. Accreditation schemes based only on building repairs do not ensure the skills needed to achieve regeneration of historic areas. The Heritage Lottery Fund does not require accreditation for the lead professionals on its own grant-aid schemes.

The current accreditation schemes are, as recognized by the sector lead bodies, based on adding a conservation overlay or 'veneer' to professionals who are already qualified in architecture, surveying, etc. This approach has minimized sensitivities among the traditional professions, but it shows no sign of delivering conservation professionals in large numbers. A more radical approach may be needed to reflect the new professional landscape and the range of potential career entry routes into conservation. The issue of levels of accreditation has also to be tackled, to encourage and develop those professionals who are interested, but may be deterred by what can appear from the outside to be a 'closed shop'.

Drivers for change?

The key potential drivers for change will remain what incentives exist to encourage or require clients to employ appropriately skilled professionals; and what encouragement is given to potential new conservation professionals. None of these are fully effective at the moment.

Incentives

Grants reach only a tiny proportion of works to historic buildings. Tax concessions for appropriate work to historic buildings would be far more effective, but the UK tax system offers incentives (VAT exemption) only for alterations, not for repairs. Furthermore these incentives are given without any check on whether the work has been carried out as approved and is of appropriate quality. The government's failure to change the VAT regime in favour of repairs has been fully criticized elsewhere, but even if these changes were introduced, the quality control issue would still have to be addressed.

Strong regulation through the listed building consent system should be an incentive for quality. Time is invested in negotiating schemes, but local authority conservation resources (constrained by government funding and performance targets) are frequently insufficient to provide effective monitoring and control of works carried out. A review of listed building enforcement is long overdue; effective sanctions are needed to raise the quality of both aspirations and outcomes.

Figure 4 Training together – a model for the future. Revival of traditional skills, with student architects and builders learning and working together on repairs to a historic building. This practical module forms part of the architecture course. Banffy Castle, Bontida, Romania.[23]

Encouragement to new professionals

Professional building conservation as a career has a very low profile. It does not feature in public sector promotion of construction careers, notably the websites of the Commission for Architecture and the Built Environment (CABE, sponsored by DCMS) and the Academy for Sustainable Communities. Conservation receives minimal coverage in architecture courses, and accreditation does not offer entry and progress for those who are interested.

Conclusion

The present situation is untenable. Changes on the scale needed can only come through positive leadership, and resourcing, from the top. The forthcoming White Paper on Heritage Protection offers an ideal opportunity for the government (and for English Heritage) to address the issues, propose solutions and implement them. Will they meet these challenges? We wait to see.

Postscript

This paper is a brief overview of a very complex situation. Its focus mainly on England is due to lack of space; the issues affect the UK as a whole. A more comprehensive version will, in due course, appear on the IHBC website.

Biography

John Preston MA, DipTP, MRTPI, IHBC, FRSA
John Preston studied architecture and art history before becoming a planner and then a conservation officer. He has been involved in conservation education at all levels, including working in schools, lecturing, organizing conferences, acting as external assessor for courses, and representing the Institute for Historic Building Conservation (IHBC) in work on standards development. He is Education Chair for the IHBC and a trustee of the Conference on Training in Architectural Conservation (COTAC). He is Historic Environment Manager for Cambridge City Council.

Notes

1 Preston, J., 'Promoting the Conservation Plan Approach', *Context* 64, December 1999, www.ihbc.org.uk/context_archive/64/plan/promote.html (accessed 20 September 2006).

2 Feilden, Sir B., 'Conservation – Is There No Limit? – A Review', *Journal of Architectural Conservation*, Vol. 1, No. 1, March 1995, pp. 5–7.

3 Feilden, Sir B., 'Architectural Conservation', *Journal of Architectural Conservation*, Vol. 5, No. 3, March 1999.

4 Orbasli, A. and Whitbourn, P., 'Professional Training and Specialisation in Conservation', *Journal of Architectural Conservation*, Vol. 8, No. 3, March 2002.

5 Maxwell, I., Heath, D., and Russell, P., 'Accreditation in Historic Building Conservation: The Work of the Edinburgh Group', *Journal of Architectural Conservation*, Vol. 10, No. 1, March 2004.

6 The Historic Buildings Council for Scotland, *1997–98 Annual Report to Parliament*.

7 *Skills Needs Analysis of the Built Heritage Sector In England*, National Heritage Training Group (2005), www.english-heritage.org.uk/upload/pdf/craft_skills_report.pdf (accessed 19 September 2006).

8 Maintain our Heritage Research Module 6: Training and Education, *Maintenance Education and Training for Listed Buildings*, De Montfort Expertise Ltd, Leicester (2003), www.maintainourheritage.co.uk/pdf/module6intro.pdf (accessed 14 September 2006).

9 Evidence to the Culture Media and Sport Select Committee's Inquiry into Protecting and Preserving the Heritage – House of Commons HC912-II 2006 – Ev 330 para 4b, www.publications.parliament.uk/pa/cm200506/cmselect/cmc-umeds/912/912we01.htm (accessed 14 September 2006).

10 Clifton-Taylor, A., *The Pattern of English Building*, Faber and Faber, London (1987).

11 Preston, J., 'A Common Grounding? – Principles, Standards, and Training', paper to 2002 Oxford Planning and the Historic Environment Conference, www.ihbc.org.uk/Papers/PATHE2002/Prestongrounding/Preston.html (accessed 14 September 2006).

12 *Review of Heritage Protection – the Way Forward*, DCMS (2004), www.culture.gov.uk/NR/rdonlyres/FCA97675-DA33-4083-9724-49432CF9 FE07/0/reviewofheritageprotection.pdf (accessed 14 September 2006).

13 *Believe or Be Left Behind*, Skills For Business (2004), www.ssda.org.uk/ PDF/Skills%20Pay%20DM.pdf (accessed 14 September 2006).

14 The new 14–19 Construction and the Built Environment Diploma is a joint initiative by 6 Sector Skills Councils, but these do not include Creative and Cultural Skills, www.cbediploma.co.uk/thepartnership/ (accessed 14 September 2006).

15 For example, www.buildingconservation.com notes a total of 47 undergraduate and 93 UK postgraduate courses (accessed 14 September 2006).

16 *Sustaining Our Living Heritage – Review of Skills and Training Needs*, Heritage Lottery Fund 2000, www.hlf.org.uk/NR/rdonlyres/AF4898F5-ADD7-4735-BC01-D2080BEC62B4/0/sustaining_heritage.pdf (accessed 19 September 2006).

17 *Sector Skills Mapping for the Heritage Sector*, Heritage Link (2006), overview at www.heritagelink.org.uk/docs/Modernising%20Sector%20Skills_final.pdf (accessed 19 September 2006), spreadsheet of initiatives at www.heritagelink. org.uk/docs/ Sector%20skills%20mapping_final.pdf (accessed 19 September 2006).

18 *Op.cit.*, Note 9, Ev 313 para 4.2.

19 *Op.cit.*, Note 9.

20 Evidence to the Culture Media and Sport Select Committee, as Note 9, Ev 204.

21 Articles accessible via '*Context* Archive' at www.ihbc.org.uk: search by subject categories 'training', education', 'accreditation' etc. (accessed 14 September 2006).

22 Feature article in *The Architects' Journal* 16.10.03; issues discussed, with links to 'before' photos, at www.c20society.org.uk/docs/casework/willowhse.html (accessed 19 September 2006).

23 See www.heritagetraining-banffycastle.org/ (accessed 19 September 2006).

What Direction for Conservation?

Some Questions

Bob Kindred

Abstract

The relationship between the government and the heritage sector over matters of legislation and policy has been characterized, since the nineteenth century, by attitudes ranging from diffidence to indifference to (at some stages) outright hostility. The present system of heritage protection has evolved imperfectly, but without evident serious difficulties – almost in spite of the government rather than because of it. Eventually, the government's decision to create an 'arm's-length' organization (quango) in English Heritage to act as its lead adviser gave the sector an independent voice, but this has proved increasingly less effective. The organization has been penalized financially for its inability to conform to the government's deregulatory and re-organizational expectations, with significant consequences for the health of the sector overall. In the last decade, much greater reliance has been placed on proceeds from the National Lottery to fund heritage projects. This funding is now under two threats. Firstly, there is a belief within government that heritage problems have now largely been solved by this largesse from gambling. Secondly, the impending scale of infrastructure investment in the 2012 London Olympics will consequently significantly reduce funding for heritage. Although this loss of resources could have been partly made good by reducing the burden of VAT, the government has allowed ill-informed prejudice to guide its policies, while its obsession with target setting for the sector, particularly the local authorities, has been misguided and unproductive.

Introduction

A perceptive reviewer in the distant future might well conclude that the legislation for the management of the historic environment in the United

Kingdom had been defined by the lexicography of conflict: battles, campaigns and wars of attrition between the government and the sector, mainly on a philosophical basis but also increasingly related to levels of investment. In retrospect, consensus over the last quarter-century will not appear to have been much in evidence; nor will the approach to heritage management have been particularly positive, inclusive or well resourced. How did this state of affairs arise, and what does it tells us about the direction of late twentieth-century conservation practice?

The legal and administrative framework of heritage management has evolved gradually and piecemeal, and, it might be argued, only with reluctance on the part of government. John Delafons, in the most authoritative guide to the politics of twentieth-century heritage policy,[1] identifies a consistent strand of reluctance on the part of national politicians and their civil servants to promote the cultural, community and civic values and benefits of the historic environment over the rights of individuals. He notes that this has been reinforced by a paternalistic, centralized approach to the identification of what should be valued and protected, if not promoted and saved from harm, for future generations to enjoy.

These attitudes extend back to the nineteenth century, as exemplified by the difficulty Sir John Lubbock MP faced in 1873 in pressing Parliament for protection of what was seen by some as 'the absurd relics' of our 'barbarian predecessors'.[2] This seems to some degree to have characterized the attitude of legislators ever since. It took nine years to get this initial, most elementary (and ineffective) statute enacted, despite criticism that it would 'take private property not for essential public purposes such as railways, "but for the purposes of sentiment", and it would be difficult to see where that would stop'.[3]

Delafons demonstrates, with stark clarity, the lack of concern for heritage both before and after the Second World War, highlighting the fact that between 1945 and 1951 the new Ministry of Town and Country Planning issued nearly a hundred circulars on the implementation of planning legislation, but none made any reference to protection or management of historic buildings. In this policy vacuum, and despite concerns about the impact of the war-time losses of ancient buildings, professional planners were given no lead on heritage matters in the context of reconstruction, and a central focus on the brave new world of urban design, garden suburbs and new towns emerged.[4]

That this philistinism extended to the highest levels of government is perhaps most graphically demonstrated by the demolition in 1962 of the Euston Arch, a huge neoclassical listed structure outside the entrance to London's first railway terminus designed by eminent architect Philip Hardwick and completed in 1858. Despite vigorous protest by the *Architectural Review* and a deputation being sent from the Royal

Academy, the Society for the Protection of Ancient Buildings (SPAB), the Georgian Group and the Victorian Society to meet with Prime Minister Harold Macmillan in person, he refused to intercede and the structure was demolished. This marked a significant turning point in attitudes to conservation outside government, but sadly not within it.

A further demonstration of this ingrained government conservatism, verging on outright hostility, was the attitude of senior civil servants when later faced with, to them, the alarming prospect not only of heritage legislation being proposed from outside government (the 1967 Civic Amenities Act), but with the potential for it to be applied to whole geographical areas, not just (at that stage a limited number of) individual buildings and monuments.

Immediately prior to the introduction of the Civic Amenities Bill, Richard Crossman – then Minister of Housing and Local Government, and a man with a particular interest in historic buildings – insisted that the listing of historic buildings should be centralized within his Department. His Permanent Secretary, Dame Evelyn Sharpe 'utterly despised' this work.[5]

> She regarded it as pure sentimentalism and called it 'preservationism', a term of abuse. She, who counted herself a modern iconoclast, took the extreme – yes I will say it – illiterate view that there was a clear-cut conflict between 'modern' planning and 'reactionary' preservation.

Crossman went on to remark that he found the actual division which dealt with historic buildings to be 'extremely rigid and difficult', and that 'they continuously resented my interference and tried to defeat me whenever they could' (when he tried to get buildings listed). 'Altogether my relations with them were very hostile', he concluded.

Occasionally, ministers were no better. Delafons recalls the *Architects' Journal* reporting in 1976 that Lady Birk, a junior minister in the Department of the Environment (DoE), had given a speech stating that 'too many buildings have been listed', and that in future 'we shall not be giving so many buildings the benefit of the doubt'. She later claimed she meant too many 'marginal' buildings – particularly those of the late nineteenth century, but this hardly offered any reassurance, and was ill judged in the light of growing interest in buildings of that period. Fortunately she was wrong, and no other minister has ventured to question the listing process or cast doubt on the government's commitment to conservation in this way. In 1977, Lady Birk proposed a slowing down of the listing resurvey of England to meet staffing costs, prompting Derek Sherborn, Principal Inspector of Historic Buildings to describe the minister as 'the most disastrous we ever had'.[6]

Civil servants and politicians regularly change, of course, and responsibility for listing of buildings has now, almost 40 years later, been finally removed from government. Although Delafons documented the eventual transfer of the wider advisory, policy and grant-making functions to English Heritage in 1984 (and with it many former civil servants), and observed the creation of a Department of National Heritage in 1992, he could hardly have anticipated the subsequent extent to which government disinterest in heritage, bordering on open hostility, has proceeded in recent years. Nor did he anticipate the diminished influence of English Heritage within and across government. No doubt Richard Crossman would have been appalled.

Government and community-led heritage

Although the present system of heritage protection, as it has evolved since the end of the Second World War, has served the sector reasonably well, it has not been without its shortcomings. It is widely recognized that it has taken insufficient account of the substantial growth in the numbers of protected heritage assets since listing and scheduling began; nor has it acknowledged the range of designation types (Parks and Gardens, Battlefields and World Heritage Sites for example). It is debatable, however, whether this is so confusing and difficult for ordinary property owners and the general public to understand that the system requires almost complete reinvention, with all the resource implications this implies.

The system has also failed to keep pace with the evolution in interest, appreciation and understanding about our historic environment, our international treaty obligations, case law and the recognition of a more holistic basis for managing it (although again, the deregulatory instincts of the government suggest neither present nor future legislation is likely to extend a holistic approach much beyond existing designations).

The public's appetite for heritage has also continued to increase, not least through the power of television with programmes based on Pevsner's Buildings of England series, *Restoration* for buildings-at-risk and *Time Team* for archaeology. Popular definitions of heritage have expanded, as has the scope to fund it through the proceeds of the National Lottery. This may have prompted thinking within government that everything is heritage to someone, but may not be an expression of popular support, nor that this might be beneficial.

Recently, questions prompted by the Heritage Lottery Fund's 2004 conference 'Who do we think we are?' asked whether the expansion of 'heritage' has resulted in historic buildings 'carrying a weight of association which can be uncomfortable or off-putting'.[7] This led a report by the

Attingham Trust, *Opening Doors*, to define 'intangible heritage' as 'whatever people like to think of themselves'.[8] It was argued that by following this trend towards an ever broadening and inclusive definition of heritage incorporating everyday experiences, the concept would become almost meaningless. Heritage as a discrete area of life would be replaced by life itself – its mission watered down. Actual places and artefacts would be abandoned in favour of what the HLF called the 'intrinsically intangible and ephemeral... the living experiences of people'. By this process, places of universal historical significance and of national value give way to the local and parochial; there must be no intellectual challenge, no effort to understand or appreciate. Thus is a patronizing undercurrent detected, and a tension exposed between the special-elitist versus the commonplace-populist.

The apparent motivation for this appeared to be the government's obsession that recipients of its funding should meet quantitative targets for new audiences for heritage (referred to below). This constituency would apparently only be won over by relating this to 'their' particular experience, with a personalized heritage being dressed up in the clothes of empowerment. If heritage continued in this vein, it was concluded, it was likely to talk itself out of existence.

Despite the government's apparent past flirtation with heritage being defined from the bottom-up, its approach has remained resolutely a top-down, 'command' process; one characterized by the present Heritage Minister, David Lammy – without any hint of irony (given the responsibility of the Department of Culture, Media and Sport, or DCMS, for heritage policy) – as 'experts talking to experts'.[9]

Opportunities to review this approach have come but seldom. A comprehensive and detailed review of the sector – co-ordinated by English Heritage in 2000 and published as *Power of Place*[10] – made 18 recommendations, but it did not recommend a complete recasting of the legislation. Nor did this emerge explicitly from the government's somewhat inadequate response published as *A Force for Our Future*.[11] It was only in November 2002 that the Secretary of State for Culture somewhat belatedly announced a review of heritage protection. This was seemingly beyond the general thrust of the sector's advice in 2000, but the announcement did coincide with a critical appraisal of English Heritage's role in that organization's Quinquennial Review.

It was stated in 2003, at the outset of what was eventually to become the government's Heritage Protection Review, that the intention was to approach issues of protection without preconception and examine the potential of a new system of designation which might recognize and give weight to what is of value to local people; however, this concept did not survive to be tested at the public consultation stage.

Had such an innovative concept emerged into the public arena, it would have been entirely novel, and would have proved contrary to both the government's centralizing economic policy and its deregulatory instincts. It was therefore odd, to say the least, that the Secretary of State for Culture, speaking at the same conference as her Heritage Minister in January 2006, should claim:[12]

Instead of funding what we think is important, we'd start by asking people what's important to them and then thinking how best to protect it. In terms of heritage this would mean asking the public which buildings and open spaces they value in their local area, and then allocating funding accordingly.

The government was 'looking at how we give ownership [of the nation's heritage] back to the local communities themselves' and the heritage sector was perceived as being 'only willing to engage with communities on their own terms – and in a language that excludes those to whom they are talking'.[13]

Existing national conservation area legislation already enables this local protection (and 'ownership' through requirements for statutory public participation and consultation); yet, local communities have been denied any other form of control over meaningful definitions of heritage. It is also questionable whether there would be any willingness on the part of the Secretary of State to persuade the government to fund such forms of un-designated heritage asset – except via the Heritage Lottery Fund. An even less likely source of funding for them would be local authorities, over whose priorities or expenditures DCMS has no direct control.

The Secretary of State's proposal would reduce heritage protection to vox pop evaluation without any standards for designation or prioritization for resources, and without any consideration of the need for an informed constituency of support. In all, this betrays what Tony Burton of the National Trust has described as: 'When asked about the historic environment, DCMS can come out with the right words in the right order but it is not part of the narrative of the Department as a whole.'[14]

Proposed future legislation will almost certainly not address the issue of prioritizing localized heritage value (except by way of the possibility of a local section of a unified national heritage register), despite lip service being paid to a holistic view of protection. While the government has made a commitment not to weaken existing protection, it has equally made no undertaking to strengthen the legislation. Given the long history of the erosion of those assets already of concern to local communities, conservation areas and areas of local significance,[15] this is a serious omission and one emphasized as a particularly pressing need by organizations ranging from the SPAB and Heritage Link to the Institute of Historic Building Conservation and the Royal Town Planning Institute.

Overall, the interest in heritage, which has grown immeasurably in the last 30 years, has usually been associated with the cherished local environments where people live and work, but not necessarily those recognized as worthy of statutory protection; even where this is the case, such areas are protected in a cursory way and are vulnerable to unsympathetic change. Furthermore, although the Heritage Lottery Fund has widened the public's perceptions of 'heritage' and the definitions of what it will support financially, in the context of the recent consultation on heritage protection,[16] these boundaries have almost certainly been extended well beyond what the government would comfortably recognize as worthy of its legislative attention. Indeed, the relegation of cherished local environments, in the form of designated conservation areas, to the local part of any future unified register of heritage assets would suggest a lack of national recognition of local community concerns, where a sense of 'ownership is most profound'.

Local communities today are faced with specific threats, possibly including: rapid urbanization; housing market renewal;[17] pressures for expedient regeneration; the pace of modern development; or simply the general often subtle but steady incremental degradation of their local environments in the form of so-called 'improvement' or 'progress'. These are carried out with precisely that kind of characteristic spirit of British individualism deployed against Sir John Lubbock in the nineteenth century, and any opposing view would probably not find sympathy within government, especially if it were based on the 'sentimentalism' deprecated by Dame Evelyn Sharpe in the twentieth century. A major challenge for the twenty-first century is to see whether such attitudes in government might be overcome and if there is a genuine desire to encourage communities to engage with, and have genuine involvement with, the management of change.

A perception has long been established within government (and seems to persist), that the historic environment is a barrier to progress, is rooted in nostalgia, is 'just about bricks and mortar' and is a constraint on development. By declining to address a truly bottom-up, community-based approach to heritage protection, the government implies the promotion of a double standard for the historic environment in terms of what is managed and how it is managed. Faced with the potential perceived impediments to economic prosperity and commercial enterprise, the instinctive reaction of government appears to be to direct policy centrally (to stimulate market forces), but nevertheless to deregulate land-use planning without the adequate safeguards while failing to trust local communities to decide what they value and wish to protect locally.

Has terminal decline set in at English Heritage?

The formation of English Heritage in 1984 raised high hopes in the sector that some degree of independence from government might enhance the status of the historic environment and lead to its greater promotion and appreciation. There is no denying that English Heritage has achieved a great deal in its 22 years of existence, has an unrivalled concentration of expertise and is well regarded throughout the sector. However, in the past decade its influence with its sponsoring department and within government as a whole appears to have been steadily decreasing as its resources in real terms have been relentlessly cut back. In 2000, the Secretary of State had to hastily deny publicly that she had no plans to abolish the organization or merge it with others.

It was perhaps a forlorn hope to expect English Heritage to be truly independent from the government when its independence did not extend to its budget. Furthermore, in the same way that successive Secretaries of State have focused on major iconic elements of the heritage, particularly those with tourism potential – 'anywhere where there is a turnstile they like it'[18] – so English Heritage has concentrated most of its activity on major highly graded historic buildings, representing only about 6% of the resource, while not being demonstrably too concerned about the care of (the vast majority of) the remainder. Only recently, after English Heritage has seen its resources seriously curtailed for the first time since its inception, has it appeared to realize with any clarity that it cannot in future operate as it once did. Consequently, it has become been more concerned about the skills and capacity of conservation services to deliver at local government level what it clearly cannot.

English Heritage is a hydra-headed animal. It has regulatory functions, administering much of the existing heritage protection regime; it acts as the government's principal adviser on policies affecting the historic environment; it is a commercial operator of visitor sites; it is a distributor of grants for repairs (particularly to places of worship and to an ever-decreasing extent to private owners or on an area management basis with local authorities); and it aims to build capacity in local authorities, the voluntary sector and others to assist in their efforts to preserve, and promote engagement in, the historic environment.

English Heritage maintains that 'there are significant benefits from [its] structure as an integrated organisation'.[19] The National Trust (and indeed the Historic Houses Association and the Country Land and Business Association), in giving evidence to the Commons Select Committee in 2006, feared a possible conflict between the various roles, warning that 'when resources are tight', English Heritage will have 'to make choices between its responsibility to look after the assets in its direct management, its role as an adviser, and its ability to deliver wider sector support'.[20]

Written submissions from a number of organizations to the Inquiry also argued that English Heritage was trying to do too many different things and that, particularly in the light of its financial situation, it should aim to 'do less but better'. This objection, though, is generally made from the standpoint that the resources are spread too thinly, rather than that there is a fear of serious conflict between the different roles.

Where English Heritage has become involved with historic environment management at local authority level it has had distinctly mixed success. In its early years its policy (following that of the government's Historic Buildings Council) and funding initiatives recognized that joint conservation work would require patient long-term commitment. Its area – based Town Schemes would often run for five years and often be extended for a further five (or more); would be administered in an unbureaucratic fashion, and would lever in many times more private regenerative investment. Later, with its funding pressures increasing and the introduction of target-driven government, its subsequent initiatives became much shorter in duration, often were not extended, and were significantly more cumbersome to apply for and operate. For most of its existence, English Heritage has also preferred to give eye-catching large grants to major higher grade buildings than to concentrate on promoting good maintenance of all listed buildings. The latter might have obviated the need for some of the grants requested, and this agenda has only been engaged with belatedly.

On other fronts where patient long-term continuous commitment was also required, such as resolving buildings-at-risk cases (estimated variously as numbering between 20,000 and 37,000 buildings), initial momentum with local authorities was not maintained, making it more difficult to re-establish the momentum later. In part, this was beyond the control of English Heritage. The government had always made clear that more central funding to tackle this problem could only be expected if the extent of the problem could be accurately quantified. By the time this had been done to any extent in conjunction with supportive local authority partners, the main policy focus and spending priorities of English Heritage had shifted elsewhere. It took a further seven or eight years to reprioritize buildings-at-risk within the organization.

While some local authorities had continuously pursued solutions for buildings-at-risk from the mid 1980s (and can now point to a significant reduction in the problem), English Heritage only returned to the issue in 1998 with its own At-Risk Register, and this only represented highly graded problem buildings (just 6% of the total resource). Almost 20 years after this issue was first highlighted, the precise extent of the problem in

England remains undefined, both in terms of numbers and the resources required to tackle it. Large areas of the country have no comprehensive and/or up-to-date surveys – let alone a strategy to resolve the problem. English Heritage's estimation of the cost of dealing with the items on its own register amounted to £400 million, an alarming figure for a government unprepared for any additional heritage expenditures. Throughout this period, the government conspicuously failed to meet its commitment in Parliament to review (and hopefully to improve) the legislative provisions relating to heritage enforcement that would have assisted the management of the issue.[21]

Why does the government see English Heritage as a problem?

A perception clearly formed within the Labour government after 1997 that English Heritage was a problem. It was seen as anti-development, which some might argue was a consequence of the proper execution of its purpose, but this had a fundamentally serious impact on its subsequent funding. In 1999 it was required to make a cut of £44 million in its budget over four years, together with a costly regionalization of its services (involving new premises and staff relocation to nine parts of the country including such expensive areas as Surrey and Cambridge). A further funding squeeze followed for the period 2000–1 to 2006–7.

This was clearly a form of government sanction when compared to the funding of the Arts, which over the same period received a 75% increase, and Sport, which obtained a 195% increase. When taken with this extended timescale and coupled with the criticism of English Heritage in the Department of Culture's Quinquennial Review in 2002, one witness remarked to the Parliamentary Select Committee in March 2006: 'It is unclear what English Heritage are being punished for.'[22]

The recent Commons Select Committee Inquiry has demonstrated that English Heritage has provided a high-quality service over the years, and has developed greatly respected expertise and shown enthusiasm; the consensus, though, is that funding pressures and structural reviews have forced it to reduce its activities to a point where its influence inside and outside government is seriously weakened. Expertise has been reduced through 'modernization' involving staff cuts and early retirements among many of its best practitioners. In evidence, Tony Burton stated 'At the end of the day, we think the cycle will come round and we will have to reinvent a better funded English Heritage as the solution to the mess we have got ourselves in.'[23]

From 2000–1, English Heritage has had a reduction in real terms of £9.7 million in what it received from the Department of Culture as its government sponsor. Worse still in terms of initiatives foregone, workload curtailed or grant recipients affected by these cuts, the settlement in the government's 2004 Comprehensive Spending Review led to a further reduction of £14.4 million for the three-year period 2005–6 to 2007–8 (seen against an overall government contribution of £127.9 million by 2007–8). Overall, this has amounted to a real reduction of £24.1 million over eight years.[24] Furthermore, income from its rural visitor sites was significantly damaged by the serious outbreak of foot and mouth disease.

Equally dramatic has been the impact on English Heritage's purchasing power, which is £12.9 million less in the five years since 2001 – and if building tender price inflation is considered, which directly relates to the cost of building works and applies to 75% of their grants – it is £27 million less in real terms over the same period.

Although English Heritage have claimed they have been able to deliver efficiency savings in the way they operate, this appears to take no account of the expected additional costs required to introduce the new Heritage Protection system in 2010, which must therefore be absorbed within existing already constrained budgets at the loss of other work.

Has the government understood the heritage impact of VAT?

The iniquitous, pernicious nature of Value Added Tax (VAT), which penalizes the repair and maintenance of historic buildings but encourages their alteration, has been one of the most damaging influences of government fiscal policy on the historic environment in recent decades. While its effects have been well documented elsewhere, what has been most disturbing is the intransigence of the government in the face of the convincing evidence of the need for change presented by the sector itself. Concern was expressed within the sector at the apparent failure of the Department of Culture to persuade the Treasury of the case that the department was vigorously acting as the government champion of heritage policy.

In 2004, the House of Commons Office of the Deputy Prime Minister (ODPM) Select Committee held an Inquiry into 'The Role of Historic Buildings in Urban Regeneration'.[25] The committee took detailed and extensive evidence from the heritage sector and the development industry. It was widely acknowledged among the former that the committee's final report was more sympathetic to, and more supportive of, the positive regenerative effects of the heritage sector than might have been expected, and highlighted the fact that VAT was a problem.[26]

A detailed case against the imposition of VAT on repairs was demonstrated, and although the committee was not one to regularly consider heritage matters under its usual remit, the report Summary stated unequivocally that:[27]

> The VAT treatment of construction work on historic buildings is perverse and a disincentive to projects involving their reuse and goes against the government's sustainability principles. New-build schemes are favoured rather than the reuse of existing buildings, and alterations are encouraged rather than repairs to them. The tax system should favour, not deter, the preservation and reuse of historic buildings.

The committee did not call for the work to be zero-rated, only that it should be reduced.

While it is acknowledged that negotiations on rates of VAT require complex and often protracted negotiations within the European Union, the response of the government in November 2004 to the Select Committee's report was astonishing, stating:[28]

> We will continue to keep the impact of VAT on different types of building work under review. However, we have seen no compelling evidence that the absence of a reduced VAT rate on repairs significantly hinders the maintenance of historic buildings, and no evidence that most of the balance of a blanket relief for repair and maintenance work would not go to middle and high income households making improvements to houses already in a good state of repair.

So while the government was content to continue to penalize individual owners, it seemed to see no contradiction in extending partial and favourable treatment to the ecclesiastical lobby by offering the VAT concession to churches.

If the government was thinking of middle- and high-income households as those represented by, for example, the Historic Houses Association and the Country Land and Business Association, it did not say so; and while both associations drew on a particular constituency of support principally from rural country houses, the government appeared oblivious to the plight of ordinary domestic owners who lack a similar focused and articulate lobby.

Of the nearly 500,000 listed buildings in England, almost 70% are residential, and most are in private occupation. There are an estimated further 4.4 million dwellings in conservation areas across the spectrum, from inner city suburbs to market towns to villages. This very considerable part of the sector is without a representative body, and these people have been generally disregarded in all the work done and all the policy documents published since *Power of Place* in 2000.[29]

In the modern era, ownership and maintenance of individual historic buildings has generally been penalized by the government since the introduction of VAT, because tax is payable on repairing historic fabric rather than its alteration. This is despite the fact that the government expects private owners to carry the current (and implied future) responsibility on behalf of society. This reflects back to a philosophical position propounded in the nineteenth century, and particularly William Morris's dictum of our being only custodians for the time being of the ancient buildings we have inherited, and not regarding ourselves as free to do with them as we please.[30] Sadly it does not support Morris' other epithet: 'Stave off repair with daily care.' The situation is exacerbated by the huge growth in protected assets in the modern era; and while Parliament could have, at any point in the past, imposed a duty on the owners of historic buildings to keep them in good repair, it has always chosen not to do so, no doubt for understandable political reasons. Conversely it could, and the sector has argued in great detail that it should, have favourably adjusted the tax regime.

The government's most recent view about partiality, made without any justifying evidence, perpetuates an unfairness within the heritage protection system inherent from the outset – namely, that there would be no compensation to owners for the communal and cultural restriction of society on an individual's freedom to change the asset. Indeed, the government's view seemed to be based on the 'extreme illiterate view' observed by Robert Crossman so many years before, and it must be assumed that the present government attitude is with the knowledge of, if not support within, the Department of Culture. This has further undermined its credibility as a champion of the historic environment.

Has governmental target setting improved heritage services?

One of the marked changes in the general approach of the present government over its predecessors – and particularly to the non-governmental organizations (NGOs) it funds, such as English Heritage – has been an obsession with detailed target setting, as set out in binding Public Service Agreements. The Department of Culture's recent target, for example, is to increase the number of people visiting designated historic environment sites by 3% by 2008, using a baseline set in July 2006. The subsets of these targets are even more specific, for example: 45% of adults from black and minority ethnic groups; 57% of adults with a limiting disability and 58% of adults from lower socio-economic groups (C2, D, Es) must make at least one visit to a historic site. Another is a requirement for English Heritage to deliver a minimum of 40 community engagement projects involving 30,000 people across the historic environment in partnership with local agencies.

The DCMS's efforts to measure participation are chiefly through its *Taking Part* survey – a continuous national household survey, which uses visits to designated historic sites as a benchmark. This approach sits oddly with the stress on public value, local buildings as heritage, and heritage being 'all around us', as foreshadowed in *A Force for our Future*,[31] and reiterated by Secretary of State Tessa Jowell and Heritage Minister David Lammy at the 'Capturing the Public Value of Heritage' conference.[32]

Other organizations, such as the National Trust and the Churches Conservation Trust, work with the Department of Culture to try to deliver these targets for visits to designated historic sites; but the National Trust has urged that it is the quality and the depth of the experience that matters and not necessarily the numbers of people and the kinds of people coming through the door.

When the Labour government came to power in 1997, it also had a remit to reform public services including local government. It did this initially by introducing the concept of 'best value' in the provision of services by requiring: comparison and external scrutiny; incremental improvement; in some cases the outsourcing of services; and regular inspection by the Audit Commission to ensure the implementation of agreed good practice models of service delivery.

Generally failing local authority services were prioritized for scrutiny, but eventually all local authority services were to be evaluated. The problems for heritage services were more the absence at the outset of service templates, with long-standing absences being more a legal requirement than a failure of the service provided.

With heritage services providing only a small part of planning services in staffing and expenditure terms, and planning proving to be a small part of a local authority's overall operations in comparison with housing, education or social services – and given the huge cost of undertaking thorough scrutiny and regular external inspections – with hindsight the Best Value initiative might have better been called Best Practice initiative, and confined to the development of model templates and the dissemination of good practice.

Quantitative improvement targets were set (and still operate) for local authority planning services (for example in the processing of planning applications). This approach proved rather more problematic for heritage services because of the qualitative nature of the outputs involved, given: the range of ways of delivering the service; differing tiers of administration providing differing types of expertise (counties, districts, unitary and metropolitan authorities); and the wide disparity of heritage resources (both staffing and heritage assets) between one authority and another. This process was not assisted by having so little of a local authority's building conservation services provided on a statutory basis, making these vulnerable to short-term, local political and financial pressures.

More recently, the government has introduced more generic testing of local authority services through a Comprehensive Performance Assessment of the authority as a whole, and these in turn are being simplified because of the huge cost of inspections running to £500 million per year.

Despite all these difficulties, the government considered that three heritage-related targets should be introduced in 2004–5 for, respectively, the management of local authority owned heritage assets; the appraisal and management of conservation areas; and the provision of Historic Environment Records. The government failed to respond to overtures from the sector, instead suggesting that a simple, effective and useful target would be to test for the regular identification of which buildings are 'at-risk' and/or the publication of a regular Register.

The three agreed targets were extensively tested by pilot local authorities in 2003–4, but only the one related to conservation area appraisal and management was finally introduced in April 2004. The other two were withdrawn at a late stage because of fears that local authorities would require additional government resources to implement them (conservation areas as an existing statutory requirement seemingly requiring no additional resources). Disquiet in the local authority heritage sector has followed because of the government's unwillingness to finance this small-scale target setting and compliance within local government – even when the proposed revisions to the heritage protection regime imply massive additional resources at both English Heritage and local authority levels.

Target setting for local authority heritage services has proved to be of very limited value given the nature and range of services being provided. Questions must be asked about the extent to which the government really understood so many issues: the nature of what local authorities provide, the relative priorities given to different aspects of the services, the nature of the relationship to building heritage and archaeology, and the seriousness of the government's intentions to make these effective without adequate resources.

Conclusion

Throughout most of the last 100 years, the government's commitment to heritage protection can be called into question. In its infancy, the heritage sector (if that is how it could be so described) relied on committed individual politicians to promote legislation that governments would later adopt, support, update and integrate. As the sector grew in size and complexity, functions were separated off into quangos like English

Heritage, but their genuine independence, particularly financial, has proved illusory, and their operations have became seriously circumscribed when they have not been able to adapt to the changing national agendas, political priorities and circumstances beyond the sector.

Government's approach to local heritage has been distinctly ambivalent, supposedly encouraging community engagement, but without empowering local communities to protect it adequately. Where government fiscal or operational initiatives originating outside the heritage sector have operated upon it, such as the impact of Value Added Tax or the general management of local authorities, the failure of government to give due consideration to their heritage implications (through its legislative and financial control over dependent local authorities) reflects a longstanding indifference to heritage, and does not augur well for future heritage reforms. These may be the subject of legislation in 2007–8, and even if they are appropriate for the needs of the sector and survive their passage through Parliament without serious amendment (and more quickly than Sir John Lubbock's Bill), it will have taken a decade from *Power of Place* to begin to reform the system – and yet, without any real sense of serious cross-government commitment to make it a serious priority – thus reflecting over a century of general disengagement.

Biography

Bob Kindred MBE, BA, IHBC, MRTPI
Bob Kindred has headed the Conservation Service in Ipswich since 1987. He was formerly editor of *Context*, the professional journal of the Institute of Historic Building Conservation, for a decade until 2000, and is a member of the Institute's Council and Policy Committee. He is one of five Heritage Advisors to the Local Government Association. He was an external advisor on the Department of Culture, Media and Sport Project Board for the Heritage Protection Review (2003–4) and Special Advisor to the Parliamentary Culture, Media and Sport Select Committee Inquiry into Heritage Reform 2006.

Notes

1 Delafons, J., *Politics and Preservation – A policy history of the built heritage 1882–1996*, Spon, London (1996).
2 Ibid, p. 25.
3 Ibid, p. 25.
4 Ibid, p. 76.
5 Crossman, R. H. S., *The Diaries of a Cabinet Minister: Vol. 1: 1964–66* (Edited by Janet Morgan), Hamish Hamilton & Jonathan Cape, London (1975), p. 331.
6 Sherborn, D., *An Inspector Recalls – Saving Our Heritage*, The Book Guild, Lewes (2003), p. 165.

7 'Who do we think we are? Heritage and Identity in the UK Today' conference organized by the Heritage Lottery Fund (HLF) at the British Museum, 13 July 2004.

8 Waterfield, G. (ed.), *Opening Doors: Learning in the Historic Built Environment*, An Attingham Trust Report (2004), www.openingdoorsreport.org.uk (accessed 28 September 2006).

9 Speech to 'Capturing the Public Value of the Heritage' Conference, Royal Geographical Society, London 25–26 January 2006, proceeding, p. 7.

10 *Power of Place: The Future of the Historic Environment*, English Heritage, London (2000).

11 *The Historic Environment: A Force for Our Future*, Department of Culture, Media and Sport and Department for Transport, Local Government and the Regions, PP378 HMSO, London (2001).

12 Ibid.

13 Ibid.

14 Transcript of Evidence by Tony Burton to the House of Commons Culture, Media and Sport (CMS) Select Committee Inquiry into Heritage Protection, Oral Session, 14 March 2006, Vol. III HC912-III, HMSO, London (2006), Ev. 19, Response to Q. 54.

15 *Townscape in Trouble*, English Historic Towns Forum, Bath (1992).

16 'Protecting our historic environment: Making the system work better', Consultation Paper, Department of Culture, Media and Sport, London (2003).

17 The Housing Market Renewal Initiative, otherwise known as 'Pathfinder' scheme, proposes demolition of between 100,000 and 400,000 pre-1919 terraced houses in the Midlands and North of England at a cost to the public sector of £1.2 billion before 2008.

18 Transcript of Evidence by Dr I. Dungavell to the House of Commons CMS Select Committee Inquiry into Heritage Protection, Oral Session, 14 February 2006, Vol. III HC912-III, HMSO, London (2006), Ev. 10, Response to Q. 32.

19 'Protecting and preserving our heritage', Written Evidence Vol. IV, House of Commons CMS Select Committee Inquiry into Heritage Protection, HC912-IV, HMSO, London (2006).

20 Ibid.

21 Robert Carwash QC comprehensively reviewed the Planning Enforcement regime in 1991, but listed building enforcement was ignored by the government on the basis that as there were now two Planning Acts, the historic environment should be considered separately. Baroness Blatch gave a clear government undertaking in the House of Lords in 1992 that the enforcement provisions of the 1990 Planning (Listed Buildings and Conservation Areas) Act would be dealt with later, but they were not.

22 Transcript of Evidence by Mr Peter Hinton to the House of Commons CMS Select Committee Inquiry into Heritage Protection, Oral Session, 28 March 2006, Vol. III HC912-III, HMSO, London (2006), Ev. 73, Response to Q. 222.

23 Ibid, Ev. 21, Response to Q. 67.

24 English Heritage Written Submission to the House of Commons CMS Select Committee Inquiry into Heritage Protection 2006.

25 'The Role of Historic Buildings in Urban Regeneration', House of Commons ODPM Select Committee, 21 July 2004, Vol.1, Report HC 47-1, HMSO, London (2004).

26 The Autumn 2004 Newsletter of the Ancient Monuments Society described the recommendations of the report as 'words of real comfort' [to the sector] and that MPs had 'lambasted the current VAT regime'. 'Many of its findings could have been written by the Conservation Movement', it concluded.

27 Ibid.

28 Government Response to ODPM Housing, Planning, Local Government and the Regions Committee Report on 'The Role of Historic Buildings in Urban Regeneration', Deputy Prime Minister and the First Secretary of State, HMSO, London, Cm 6420, November 2004.

29 *Power of Place: The Future of the Historic Environment*, English Heritage, London (2000).

30 Morris, W., *Manifesto*, The Society for the Protection of Ancient Buildings (1877).

31 *The Historic Environment: A Force for Our Future*, Department of Culture, Media and Sport and Department for Transport, Local Government and the Regions, PP378 HMSO, London (2001).

32 'Capturing the Public Value of Heritage' conference was held at the Royal Geographical Society on 25–26 January 2006. The conference was sponsored by the Department for Culture, Media and Sport, English Heritage, the Heritage Lottery Fund and the National Trust.

The American Contrast

Donovan D. Rypkema

Abstract

Advocates for historic preservation in the United States have always looked to Europe in general and the United Kingdom in particular for guidance in conserving our valuable, albeit young, architectural heritage. The goals of US preservationists are substantially the same as those in the UK – to identify, protect and enhance our historic built environment. But the means of achieving those goals are decidedly different. To put it in oversimplified terms: in the UK, the process is largely national, top-down, regulation-driven and carried out by the public sector; in the US, the process is largely local, bottom-up, incentive-driven and carried out by the private sector. These contrasts are driven by multiple causes, but they fundamentally reflect the US commitment to the rights of ownership versus the British commitment to the responsibilities of stewardship. The US approach is neither better nor worse, but it has evolved to respond to the political and philosophical realities of the United States.

Introduction

Two hundred and thirty years ago, the American colonies declared their independence from Great Britain, but most facets of American life have never left England. The language, laws, culture and economy of the US are all firmly rooted in their English heritage.

But it is the US adaptation (which some might call mutation) of these English precedents that has left us cousins rather than brothers.

This evolution has affected not just law, language and economy, however; it has manifested itself in our different approaches to protecting the historic built environment as well. While these differences are multitudinous, this article will focus on the divergence between the US and the British approaches in three broad areas: the level of governmental regulation; the role of incentives; and the substantial differences in the roles of the National Trusts. The three resulting perspectives will correspondingly reflect the public, private and non-profit sectors of the two systems. It will conclude with some observations on weaknesses in the US system.

Even before we consider the differences in substance, however, a difference in vocabulary arises. In the United States, the movement to identify, protect and enhance the historic built environment is called 'historic preservation' rather than 'heritage conservation', as it is in most of the rest of the world. Advocates in our movement are called 'preservationists'. 'Conservation' and 'conservationists' are more commonly associated with the natural environment. When the term 'conservation' is applied in the built heritage field, it generally refers to the physical conservation of materials, whereas 'preservation' has broader political, policy and social implications.

While one might argue that this is only a semantic difference, it is unfortunate that the US vocabulary has evolved to use 'preservation' rather than 'conservation'. One primary entry in the *American Heritage Dictionary of the English Language* defines 'preserve' as: 'to keep in perfect or unaltered condition: maintain in an unchanged form'. That is certainly not the result (and rarely the goal) of the US preservation movement.

The levels of government regulation

The national role

The United States, as a nation, was originally conceived as a federation of individual, sovereign states. In spite of massive changes in the political, cultural and economic environment over the past 260-odd years, there are still many remnants of this federalist philosophy. The recognition and protection of historic properties is an example.

There are official listings of US heritage properties on the national level. The National Park Service, a branch of the US Department of the Interior, is the federal agency responsible for maintaining the inventory of these properties. This inventory is called the National Register of Historic Places. There are approximately 79,000 listings on the National Register. A listing can be either an individual building or a historic district. The highest category is a National Historic Landmark (NHL). There are less than 2,500 NHLs, and virtually all are so identified because of their national significance.

Because, however, a 'listing' is often a historic district, the total number of resources on the National Register is nearly 1.2 million buildings, sites and structures. A building can either be individually listed on the National Register or be identified as a contributing building in a National Register district. While designations as National Historic Landmarks are limited to resources of national significance, the vast majority of other National Register properties are listed because of their state or local significance.

So there is an extensive inventory of the US's historic resources. What protection does such a listing provide? Almost none. On the federal level, there is no provision that would preclude an individual owner of a National Register property – even a National Historic Landmark – from tearing it down tomorrow.

There is one exception. If a federal government agency is undertaking an action that might affect a National Register property, whether or not the property is government owned, the agency has to determine if the proposed action might have an adverse impact on the historic resource; if this is deemed to be the case, it has to try to identify alternatives, or otherwise mitigate the adverse impact. This requirement also exists for properties that have been identified as eligible for National Register listing, but are not yet enrolled. As part of that assessment, the federal agency is required to consult with the State Historic Preservation Officer (SHPO) and, if appropriate, to hold public hearings. If no resolution can be achieved on the State level, an independent federal agency – the Advisory Council on Historic Preservation – will become involved to attempt to find an acceptable solution. This provision affects all federal agencies including the Department of Defense. (There are some 15,000 historic properties on active Army bases, for example.)

This requirement was enacted as part of the National Historic Preservation Act of 1966. This provision is Section 106 of the law, so this process has come to be known as '106 Review'.

Essentially the same procedure is required when federal monies are being spent by a lower level of government. For example, state highway departments receive most of their money from federal highway funds. Therefore, if a state highway department uses those federal funds to widen a bridge, for example, it would have to measure what impact, if any, that action would have on nearby National Register properties. If potential adverse impacts are identified, there is a mandatory consultation with the appropriate state officials. After such consultation, the highway department would be expected to either change their plans or develop a mitigation proposal.

But take federal funds out of the picture and what protections does a listing on the National Register provide? Virtually none. Go ahead, tear it down. The federal government cannot stop you.

The state role

So if the protection for historic properties is weak at the national level, what about the state level? Alas, in most cases it is even weaker. Every state has a designated State Historic Preservation Officer (SHPO). While this office is a part of state government, most of its funding comes from the national government, and it correspondingly has some authority delegated from the National Park Service. As noted above, the SHPO has the central role in a 106 Review. Nominations for the National Register and application for tax credits in each state, discussed below, also flow through the relevant SHPO.

Beyond those basic functions of the SHPO, however, there is great disparity among the states in terms of historic property protections. Some states have a State Register of Historic Places, but few of these have corresponding limitations on what can be done to those properties. A few states have an equivalent to the 106 Review process, which mandates similar requirements whenever state tax dollars are spent. But which of the 50 States has strong legislation for the protection and preservation of privately owned historic buildings? None of them does.

The local role

By default, therefore, any significant regulation affecting historic properties is relegated to the local level. But this is also true of land use regulations in general. There is no national land use policy, let alone a national zoning law, in the US. Very few states have meaningful land use laws. What limitations there are on how land can be used – through comprehensive plans, zoning laws or subdivision ordinances – are almost exclusively within the purview of local government.

In most cases, the regulation of historic properties is seen as part of the planning and zoning process at the local level. In the United States, this is typically done through the establishment of a local preservation commission by the city council. Generally, this body will be separate from the local planning commission, but will often share staff with the planning department. In some instances, the local planning commission will have the authority to review and set aside the decisions of the historic district commission. In other cases, any appeal of the decision of the preservation commission will have to go to the city council. In a few local ordinances, the only remedy a property owner has to the ruling of the preservation commission is in the courts.

There are countervailing ramifications of entrusting the protection of historic resources to the lowest level of government. On the positive side, it is arguably true that local citizens are the best qualified to judge what the

Figure 1 The National Trust headquarters in Washington is a National Historic Landmark, but that designation provides almost no protection. Virtually all meaningful protections for heritage buildings in the United States come from ordinances enacted at the local level.

important heritage assets are and how they can best be protected, treated, redeveloped and used.

But a downside of this system is that there are many very weak local ordinances, some providing virtually no protection; and, of course, hundreds of cities have no local preservation laws at all. In the United States, there are about 19,500 municipal governments, 6,500 of which have populations of more than 2,500 people. The total number of local preservation commissions, however, is around 2,600, so historic properties in the majority of US towns and cities are without any form of protection whatsoever.

This 'bottom-up' approach to historic preservation also means that there is considerable inconsistency across localities in how their resources are treated. There is also the constant risk that an 'anti-preservation' city council could be elected and the existing local historic district regulations repealed. This is rare, but not unheard of in the United States. It is also not uncommon that a new council will weaken an existing ordinance.

Some cities, of course, have very strong local preservation agencies – the New York City Landmarks Commission being the most notable example. But even in New York, there has been significant recent activity on the part of city council members to curb the power of the Landmarks Commission.

In sum, national level protection of historic resources in the United States is very weak, and state level protection is even weaker. Meaningful protection, where it exists, is on the local level, and is never completely free of risk from the whims of local politicians.

Incentives and the role of the private sector

There was once a debate in the English Parliament where one MP was described as 'wielding his stick in a carrot-free environment'. When it comes to historic preservation in the United States, the opposite is often true – wielding carrots in a stick-free environment. The United States falls behind the UK (and most of Western Europe) in the regulatory protection of heritage resources. To compensate, however, we have developed numerous incentives at all three levels of government to encourage the private sector to invest in historic buildings.

It is a widely held political belief in the United States – on both the right and the left – that where possible, private capital should be encouraged to be invested in areas deemed to be in the public good. Hence we have a federal tax code overflowing with provisions specifically aimed at encouraging individuals and businesses to invest their own money in such divergent areas as low-income housing, oil exploration, child care, home ownership, research and development, and historic preservation. Thus the national government subsidizes indirectly – through what are called 'tax expenditures' – what many other governments in the world subsidize directly through appropriations.

The Historic Rehabilitation Tax Credit

The most potent of these incentives is provided at the national level and is known as the Historic Rehabilitation Tax Credit. This credit has been available in some form since 1976 (its passage being attributable at least in part to the celebration of the US's bicentennial). The credit is available to individuals, partnerships and corporations for projects that meet four tests.

First, the property must be a 'historic building' which is defined in the tax code as a building that is listed on the National Register of Historic Places. This can be either an individually listed property or a contributing property within a National Register historic district. A property can also qualify as a 'historic building' if it has been determined to be eligible for listing on the National Register. In this case, the process of listing and of rehabilitating the building can proceed simultaneously.

Second, the property must fit in one of two categories in the tax code: an investment property, or a property held for use in trade or business. If

Figure 2 Four-storey warehouse building converted to health club. Most historic preservation in the United States involves the adaptive reuse of vernacular buildings rather than the restoration of monuments.

a person owns an office building and rents it out, for example, that is an investment property. Alternatively, if someone is a lawyer and owns the building housing their law firm, that is a trade or business property. A barn on a farm, an industrial warehouse or a gas station, as examples, would each fall under one or the other of these categories. A church would not qualify; nor would one's personal residence.

The third requirement is to meet the 'substantial rehabilitation' test. Substantial rehabilitation is the greater of $5,000 or the basis in the building. The 'basis' is the purchase price of the property less the amount of that purchase price attributable to land. This will decline over time with the depreciation that has been taken for tax purposes, and will increase when capital improvements are made.

The final test is that the building must be rehabilitated in a manner consistent with the 'Secretary of the Interior's Standards for Rehabilitation'. These 'Secretary's Standards' serve in the United States as the framework for appropriate rehabilitation. It is important to note, however, that these are called the standards for rehabilitation, and not preservation (and certainly not restoration). The whole premise behind the standards is to encourage the rehabilitation and adaptive reuse of buildings, not the creation of museums.

The plans for the rehabilitation will go to the SHPO for initial approval then to the National Park Service in Washington for final approval. The project will be reviewed again after completion to receive formal certification, as a prerequisite for receiving the tax credits.

If all four of the tests are passed, the investor will receive a tax credit equivalent to 20 per cent of the qualifying rehabilitation expenditures. A tax credit is a dollar for dollar reduction in income taxes payable. Therefore a project with qualifying expenditures of $1,000,000 would earn for its investors tax credits of $200,000 – essentially reducing their federal tax liability by that amount. Qualifying expenditures include all hard costs and most soft costs, but not any costs incurred for acquisition or site improvements.

As might be expected, this tax incentive does much to encourage private investment in historic buildings in the United States. In the 30 years since historic tax credits have been available, nearly $37 billion has been invested in 33,885 historic US buildings.

How tax-sensitive these investments are, however, can be seen in Figure 3. The high point in numbers of projects undertaken was reached in the mid 1980s, with a dramatic decline in evidence beginning in 1987. Why? Because in 1985, the federal tax code was completely rewritten by Congress. What were deemed at the time to be rather minor adjustments to the Historic Tax Credit proved to be otherwise. The Treasury Department estimated that the changes would cause a 10% drop in historic preservation activity; instead the decline was nearly 70%. Twenty years later, there are still only around a third as many projects annually as compared to preservation's high point in 1984.

When measured in dollars rather than numbers of projects, preservation comes out somewhat better. By the year 2000, the amount invested in these projects reached the high of the mid 1980s, and over the past five years has averaged more than $3 billion annually.

Although the programme is occasionally criticized in the US as being only useful to large projects, the data indicate otherwise. Over the past five years, roughly a third of the projects were for less than $250,000, a third between $250,000 and $1 million and a third more than $1 million.

The emphasis that the Rehabilitation Tax Credit places on adaptive reuse is also evident in the ultimate uses of the certified projects. Over the life of the tax credit programme, around 192,000 housing units were rehabilitated – the majority of these in multi-family complexes. Over that same period, however, nearly 156,000 housing units were created, meaning these housing units were constructed in what had been non-residential structures – office buildings, warehouses, department stores, etc. In the

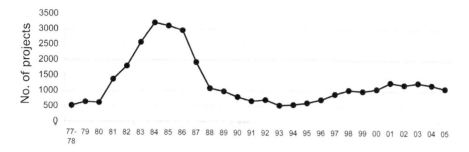

Figure 3 Projects using the Rehabilitation Tax Credit.

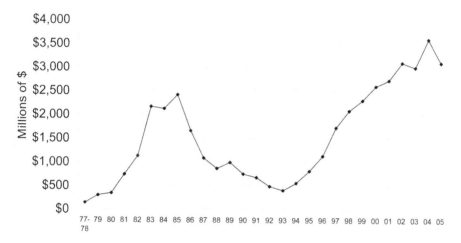

Figure 4 Investment through the Rehabilitation Tax Credit.

most recent year, nearly two-thirds of such housing units were created rather than rehabilitated.

It is also a political selling point of the tax credits that over the years around a quarter of the housing units – both rehabilitated and created – were targeted to low- and moderate-income tenants.

State and local tax credits

Incentives for historic preservation are not limited to the federal level. More than 30 states have incentives for rehabilitation using a state income tax credit, property tax relief, or both. Whereas the federal tax credit is not available for one's personal residence, most state incentives do apply to that type of property.

In part, it has been the availability of these state incentives that has generated the increase in total preservation investment in recent years. In 2004, nearly a third of all projects that took advantage of the federal tax credit also used a state income tax credit – enhancing the investors' return all the more.

Property tax incentives at the state and local level are the most common inducement. While their specifics vary widely, there is a typical model. When a historic building is rehabilitated, the property tax assessment will remain at the pre-rehabilitation value of the property for an extended period – up to ten years in some states. In simple terms, this means the following. Suppose, for example, that an owner has a property that has a value of, say, $100,000 and has a property tax levy accordingly. In the US, annual property taxes are typically between 1.5% and 2.5% of the value of the property, so in the example, between $1,500 and $2,500 per year. Now an investment is made of $900,000 for rehabilitating the historic building. Instead of the assessed value then jumping to $1,000,000, however (with the taxes jumping to $15,000 to $25,000 per year), the owner still pays taxes based on the $100,000 assessment for the number of years specified.

Like the income tax credits, this type of incentive subsidizes historic preservation through foregone taxes rather than direct appropriation. Although theoretically the same result could be achieved by simply granting the property owner a sum equivalent to his tax benefit, it is generally more politically palatable in the US to use this indirect method.

A few states, however, do have earmarked funds with which to make direct financial grants for the preservation of historic properties. The awards are competitive in nature, and there are always far more applicants (and worthy projects) than money available.

Earlier it was mentioned that in the US the 'carrot' of incentives is preferred over the 'stick' of regulations. However, nearly all of the incentive programmes will require that the rehabilitation work be completed to the 'Secretary's Standards'. So the message becomes: 'You can do whatever you like with your historic building. But if you want to use our incentives you have to follow good preservation practices.'

Some disadvantages of a private sector approach

As we have seen, historic preservation in the United States is primarily carried out with private capital, though often using publicly provided incentives. But there is another adverse consequence of this approach. As in the UK, many of the most important historic properties are owned either by some level of government (state capitol buildings, city halls, public universities, secondary schools) or the non-profit sector (churches, synagogues, fraternal organizations, private universities). Because none of these entities pay income taxes, and in most cases don't pay property taxes, a tax incentive-based preservation policy precludes them from taking fullest advantage. Public and non-profit organizations would have more resources if the general approach were direct appropriations rather than tax expenditures.

Partly to respond to this disadvantage, in recent years there has been sizable growth in non-profit participation in historic rehabilitation projects. Though these are typically called public–private partnerships, they could more accurately be described as public–private–non-profit partnerships. The statewide advocacy group Preservation North Carolina has been directly involved in two major redevelopment projects of obsolete textile mill villages. The US National Trust has established a for-profit subsidiary so that it too can reap the financial rewards of historic rehabilitation and the available tax credits. Increasingly, local preservation groups are minority stakeholders in major preservation projects. Currently, several states are examining their own inventory of historic but under-utilized public buildings to determine if there are partnership opportunities with the private sector. When the historic rehabilitation tax credit is combined with tax credits for low-income housing, a non-profit organization is nearly always involved in the transaction.

Some countries have determined, probably correctly, that from a public resource perspective the direct appropriation of funds is more efficient than the indirect method of incentives. But the economic and political culture in the United States is far more accepting of incentivizing than of appropriating.

The National Trusts

So there is clearly a difference between the United States and England in the role of the government and in the emphasis on the private sector. But those differences are also evident in the third sector, as represented by the leading heritage organizations in each country – the respective National Trusts.

Like other parts of the American experiment, the US National Trust was very consciously modelled on its English predecessor. The National Trust in England was founded in 1895 'to act as a guardian for the nation in the acquisition and protection of threatened coastline, countryside and buildings'.

Half a century later, in 1949, the US version was created for exactly the same reason, as 'an organization whose primary purpose would be the acquisition and administration of historic sites'.

Today the National Trust of England, Wales and Northern Ireland retains that original focus. It maintains over 50,000 buildings and structures of various types as well as gardens, coastlines, and 250,000 hectares of countryside.

The mission statement of the organization reads: 'The National Trust works to preserve and protect the coastline, countryside and buildings of England, Wales and Northern Ireland.'

The National Trust in the United States has evolved significantly from its original purpose, as clearly evidenced in its current mission statement: 'The National Trust provides leadership, education, advocacy and resources to save the US's diverse historic places and revitalize our communities.'

The US National Trust still owns and maintains some 25 historic properties, but that is no longer the primary focus of the Trust's activities. Again, the differences between the two organizations reflect cultural, political and economic differences between the countries.

The US National Trust has become primarily an educational and advocacy organization. If the US National Trust were being founded today, it is unlikely that it would have the direct ownership of historic properties as component of its mission at all.

It has been said that in the US, all politics is local. The historic preservation movement in the United States has learned that all (or nearly all) preservation is also local.

Thus in recent years, a major emphasis of the US National Trust has been to train, fund and mentor state and local preservation organizations. It has also been at the forefront of the 'Smart Growth'[1] anti-urban sprawl movement, and is currently leading efforts to utilize older and historic housing stock as a central strategy in response to the US's housing affordability crisis. Programmes such as 'Main Street'[2] – promoting commercial revitalization in the context of historic preservation – have made preservation advocates of bankers, city managers and business owners who haven't the slightest interest in visiting the mansions of industrialists of the Gilded Age.[3]

The National Trust in England, Wales and Northern Ireland now has 3.4 million members, reflecting the concern for and interest in the historic built environment. Indeed, the largest single source of income for the organization is memberships, which provides nearly 30% of its annual budget. Its total annual budget is over ten times that of its US cousin. As a comparison, income from memberships – of which there are only 250,000 – constitutes barely 5 percent of the US National Trust's budget.

Here are two organizations with the same name and founded for exactly the same purpose. Each would be regarded in their country as the leading voice for the heritage conservation movement. But their respective roles today are decidedly divergent.

Figure 5 Preservation-based commercial revitalization efforts such as 'Main Street' have been a major success story in heritage conservation in the United States.

Weaknesses of the US system

An objective analysis of the US system of heritage conservation would probably conclude that there are four major weaknesses:

1. With virtually no regulatory controls at the national level, there is no set of national priorities as to what will be saved or where it is most important to channel financial resources. A National Historic Landmark could be easily razed because of an absence of local protection in one location, while a property of only nominal historic value might be preserved elsewhere because of strong regulations at the municipal level. As this paper is being written, Washington University, a private educational institution in Saint Louis, Missouri, has decided to raze a National Historic Landmark building in order to build a 530-space parking garage. While preservation advocates have raised vociferous objections, there is nothing in Federal or State law that prevents them from doing so.

2. As was suggested earlier, because heritage protections are overwhelmingly local, they are also very vulnerable to the vagaries of municipal elections. In Hudson, New York, barely a year after a preservation ordinance was adopted, the City Council revised it to give themselves,

rather than the preservation commission, the power to designate properties. In Longmont, Colorado, after 34 years with a preservation ordinance, a new city council amended it to require owner consent for historic designation.

3. Because policy makers at the national level will claim, 'We are assisting historic preservation through tax incentives,' the amount of direct funding from the Federal budget is extremely small. This is true not only for monies that might be appropriated for properties not owned by the national government, but also for those for which it has direct responsibility. Many historic buildings owned by the national government in parks and on military installations are deteriorating because of years of under funding resulting in serious deferred maintenance.

4. Since preservation is predominantly a private-sector and largely tax-incentive driven activity, dollars going into historic buildings are often influenced by factors independent of the underlying importance of the structure itself. This also means that this type of investment can also be extremely volatile and will shift based on current tax rates, alternative tax preference options, interest rates, and market forces unrelated to heritage conservation.

Conclusion

The goals of US preservationists are substantially the same as heritage advocates in the United Kingdom – to identify, protect and enhance our historic built environment. But the means of achieving those goals are vastly different.

To put it in oversimplified terms: in the UK, the process is largely national, top-down, regulation-driven, and carried out by the public sector; in the US, the process is largely local, bottom-up, incentive-driven and carried out by the private sector.

These contrasts are driven by multiple causes, but they fundamentally reflect the US commitment to the rights of ownership versus the UK commitment to the responsibilities of stewardship.

The US approach is neither better nor worse, but has evolved to respond to the political and philosophical realities of the United States.

Biography

Donovan Rypkema
Donovan Rypkema is principal of PlaceEconomics, an economic development-consulting firm specializing in services to clients dealing with city centre revitalization and the reuse of historic structures. In 2004, Rypkema established Heritage Strategies International to provide similar services to worldwide clients. He is author of *The Economics of Historic Preservation: A Community Leader's Guide*, a lecturer on preservation economics at the University of Pennsylvania, and serves on the Board of Directors of Global Urban Development.

Notes

1 'Smart Growth' is the term for the broad-based sustainable development movement in the United States. It is actively supported by environmentalists, preservationists, urban planners, many politicians across the ideological spectrum, and at least some departments within the national government. Preservationists make the argument that if a city did nothing but protect its historic neighbourhoods, all of the principles of 'Smart Growth' would be automatically met.
2 'Main Street' is a 25-year-old programme of the National Trust for Historic Preservation. It began as an effort to assist the downtowns of small towns to utilize their historic buildings as part of a comprehensive downtown revitalization effort. It has expanded into neighbourhood commercial areas of large cities. Some 1,700 communities in all 50 states have had 'Main Street' programmes. It is today recognized as, by far, the most cost effective programme of economic development of any kind.
3 The Gilded Age refers to that period in US history between the Civil War (1861–1865) and the beginning of the twentieth century. The phrase was coined by the humorist Mark Twain and referred to the rich industrialists whose rapid increase in wealth was often manifested in their 'American castles', a scale of splendour and ostentation previously unknown in the United States.

All Rosy in the Garden?

The Protection of Historic Parks and Gardens

David Lambert and Jonathan Lovie

Abstract

In many ways, historic parks and gardens have thrived in the last 20 years, following a dawning of recognition heralded by the impact of the 1987 hurricanes. The English Heritage Register is now a material consideration in planning decisions, local authorities have included it in the development planning and control processes; the voluntary sector is thriving as county gardens trusts blossom, while the Garden History Society is now a statutory consultee on planning applications and a well-established part of the national amenity societies scene. High-profile restoration projects and the generosity of the Heritage Lottery Fund (HLF) towards urban parks have raised public awareness to a welcome degree.

But that is not the whole story. English Heritage is cutting resources, the HLF is tightening its belt, the National Trust is struggling to maintain standards – and in all these cases, parks and gardens tend to bear the brunt. At a local level, planning authorities, still hamstrung by the dearth of conservation expertise, struggle to resist harmful development proposals, or even to identify their stock of historic parks and gardens. After successful planning battles which established the sensitivity of these sites in the 1980s and 1990s, the tide of big and damaging leisure proposals is rising again. This paper, by the former and current conservation officers of the Garden History Society, reviews progress in garden conservation since the 1960s and offers a snapshot of the present situation.

Prehistory

Historic parks and gardens are latecomers to the feast of heritage legislation and protection. Until the 1960s, landscape parks were seen as picturesque, often over-mature, backdrops to country houses, while public parks were tarred with the same brush as Victorian architecture – blowsy, vulgar, commonplace. London squares were dusty, secluded places known only to a few keyholders. Our great Victorian cemeteries quietly surrendered their order to the encroachment of brambles and sycamores.

The historic interest of parks and gardens was similarly little recognized. Christopher Hussey's *The Picturesque: Studies in a Point of View*, published in 1927 included a chapter on gardens,[1] and H. F. Clark's *The English Landscape Garden* had been published in 1948,[2] but, although Dorothy Stroud published her scholarly monograph on Capability Brown in 1950,[3] it was not until Miles Hadfield's *A History of British Gardening* was published in 1960 that the first authoritative survey was produced. Tellingly, however, despite its immense breadth of historical research, the first edition appeared under the nondescript title *Gardening in Britain*.[4] Hadfield's was inevitably a partial view, strong on horticulture but with little to say, for example, on public parks. Similarly shy of the claim for historic interest was the extremely learned *Shell Book of Gardens* edited by Peter Hunt and published in 1964.[5] Christopher Hussey's *English Gardens and Landscapes 1700–1750* was published in 1967, and was the first work to make claims linking garden design to the great sweeps of architectural and wider cultural history.[6]

With the formation in 1965 of the Garden History Society (GHS) to promote the study of historic parks and gardens, publications started to flow. Much of the best academic work, such as Roy Strong's *The Renaissance Garden in England* (1979),[7] David Jacques's *Georgian Gardens: The Reign of Nature* (1983)[8] and Brent Elliott's *Victorian Gardens* (1986),[9] was occupied with establishing the validity of garden history as a branch of art history, relating landscape to the chronology established in the older discipline. It was not, for instance, until 1991 that the first academic treatment of Victorian public park design, Hazel Conway's *People's Parks*, was published,[10] and the historic interest of other types of designed landscapes – cemeteries, hospital grounds, allotments, town walks, squares and other communal gardens, private urban gardens, post-war designed landscapes – is only now emerging. Tom Williamson's *Polite Landscapes: Gardens and Society in Eighteenth-Century England* (1995) was the first to approach garden history from a critically theoretical point of view.[11]

Garden visiting seems always to have been a popular recreation, and many private estates had, and still have, a tradition of public opening. However, it was only with the acquisition of the ruinous Westbury Court garden in Gloucestershire in 1967 that the National Trust, for example, began to explore the idea of 'restoring' a garden to a known earlier state, as if it were indeed a work of art. This fundamentally changed the view of gardens. The restoration of Painshill began in the early 1980s, and the storms of 1987 gave a tremendous boost to the art and science of repairing gardens to conservation standards with grant-aid programmes from both English Heritage (EH) and the Countryside Commission. Heligan, in the 1990s, made garden restoration romantic and sexy, but it was arguably the HLF, and in particular its Urban Parks Programme from 1996, that finally established historic parks as mainstream heritage.

In tandem with the growth of both academic garden history and garden restoration, the GHS was drawn into the planning arena, seeking to protect sites from harmful development on the basis of their historic landscape importance. Its first appearance at a public inquiry was in 1971, arguing successfully for a diversion of the proposed route of the new A590 in Cumbria from its planned route through the park at Levens Hall. Through the 1970s, under the secretaryship of Mavis Batey, it increased its activity in this area, making common cause with other campaigning groups such as the Georgian Society, the Victorian Society, the Campaign to Protect Rural England (CPRE) and Save Britain's Heritage (SAVE).[12]

The GHS also began a lengthy lobbying process with regard to the drawing up of a list of historic parks and gardens, first proposing a 'Register of Gardens' in a newsletter in 1969. In 1973, the Historic Buildings Council's annual report referred for the first time to the importance of the landscape setting of historic buildings being in many cases:[13]

> ... of outstanding interest in its own right; indeed the 'English' park or garden as exemplified by the creations of Capability Brown or Repton is of international importance or influence. But so many such landscapes have been lost over the years that it is essential to protect and restore the remainder.

The Register of Parks and Gardens of Special Historic Interest

The lobbying of the HBC would eventually lead to the inclusion in the 1983 Heritage Act of the legislation enabling such a *Register* to be produced by the reformed Historic Buildings and Monuments Commission, English Heritage (EH). EH began work on its *Register of Parks and Gardens of Special Historic Interest* in 1984. The Act allows EH to compile 'a register of gardens and other land... appearing to them to be of special historic interest'. This was a wide brief, but until work began on the separate register of battlefields, published in 1994, EH restricted itself to parks and gardens.

The *Register* was completed in 1988, an extraordinary achievement by one man, the EH Gardens Inspector, Christopher Thacker. Working from

GHS files which had been prepared for the first nine counties, first edition Ordnance Survey maps and aerial photographs, and his own extensive knowledge, he compiled a list of some 1,000 sites for inclusion. These initial sites were in the main landscape parks and the gardens of large houses.

The first edition of the *Register* was sent to all owners and planning authorities, but it is clear that many of the latter had little idea what to do with the new information. A number took the hint and began producing their own county lists of parks and gardens; Avon was an early example, as was Hampshire. Others, such as Cambridgeshire and Norfolk also produced schedules, while a number of authorities, such as Sheffield, the Vale of the White Horse and East Sussex, produced supplementary planning guidance for historic parks and gardens.

It was not until 1994, with the publication of the government's Planning Policy Guidance Note 15, *Planning and the Historic Environment* (PPG15), that there was any clear direction given by central government on what to do with the *Register*. PPG15 advised that 'local planning authorities should protect registered parks and gardens in preparing development plans and in determining planning applications', and it advised that, while:[14]

> ... no additional statutory controls follow from the inclusion of a site on the ... Register... The effect of proposed development on a registered park or garden or its setting is a material consideration in the determination of a planning application.

Upgrading of the *Register* began in 1996 and was finally completed in 2004, greatly extending the earlier descriptions and supplying maps showing the boundaries of the registered land. More significantly from a historian's perspective, it greatly extended the understanding of historic parks and gardens; the number of nineteenth-century public parks was substantially increased, four historic allotment sites were added, along with cemeteries, hospital and asylum grounds and twentieth-century designed landscapes. Spot-registering had already plugged many of the gaps, but the upgrade and a number of theme studies allowed the *Register* to grow significantly in its definition of what constituted a historic garden. *Register* descriptions now often run to four or more pages, and have proved influential on present thinking about area designations under the Heritage Protection Review (HPR).

Since regionalization, EH has employed a number of landscape architects to contribute to their team responses on planning applications and, although several regions still do not have their own landscape architect, they are often able to contribute significantly to EH responses.

Understanding of the range and variety of typologies has grown rapidly in the wake of the publication of the *Register*. EH initially responded with spot-registering and by commissioning a number of theme studies to investigate those typologies – for example on public parks, detached town gardens for rent (allotments), garden squares and villa gardens – all of which led to new additions to the original *Register*. Studies on hospital and asylum grounds, cemeteries and post-war landscapes and gardens have also resulted in new additions to the *Register*.

Perhaps the most important growth in understanding has been in the local perspective, largely developed by the county gardens trusts movement. The preparation of local lists and schedules has been a project for many of the trusts, often initially in partnership with the local authorities and subsequently adopted in the local plan. Because the *Register* is such an exclusive list, parks and gardens have been in the forefront of promoting the local perspective on significance. Well in advance of *Power of Place*,[15] although inspired by EH's pioneering but under exposed *Sustaining the Historic Environment* (1997),[16] the GHS was arguing for a more pluralist understanding of the nature of 'heritage'.

One typological group which remains problematic is vernacular gardens, the gardens of modest, often suburban, houses. These have come to the fore recently with the London Assembly Environment Committee's report, *Crazy Paving* (September 2005),[17] on the paving over of front gardens to form car-parking areas, and the ten-minute rule Bill introduced by Greg Clarke in February 2006, which sought to address the anomalous status of private gardens as previously developed or brownfield land which leaves them easy prey to infill development. The GLA report's recommendation that the importance and protection of ordinary front gardens be addressed in the revised London Plan is an exciting one, and although at the time of writing the Bill looks unlikely to pass into law, it has served a useful purpose in raising the matter up the political agenda.

Despite all the work undertaken by EH and others in identifying sites which may be appropriate for addition to the *Register*, obviously registrable sites continue to slip through the net, only to emerge as being of special historic significance once the planning process is well advanced. A recent example of this is Briggens in Hertfordshire, where a proposed substantial hotel extension would, in the view of the Ancient Monuments Society, the Georgian Group and the GHS, inflict considerable harm on an unregistered, but eminently registrable early eighteenth-century landscape forming the setting for a Grade II listed mansion (Figure 1). The design of the landscape can be attributed beyond reasonable doubt to the Royal Gardener Charles Bridgeman (d 1738), but somehow, despite the survival of a substantial circular basin, remnants of an earthwork amphitheatre, a canal, a walled garden and elements of formal tree planting, the significance of the landscape had not been recognized. In the past, when faced with such a situation, it was a relatively easy process to seek to have

Figure 1 An example of a historic garden of registrable quality still unregistered: Briggens, Hertfordshire. (Georgian Group)

a site spot-registered; however, since the beginning of the HPR, spot-registering has been in abeyance and only now, some two years later, are the beginnings of a system emerging.

Protection

Ever since parks and gardens were first considered by the Historic Buildings Council, the possibility of their being protected in law from damaging development has been discussed. Prior to the current Heritage Protection Review (HPR), a good deal of work has been done on possible forms of registered garden consent.[18] Successive governments and their agencies have however fought shy of introducing a new level of statutory control into the field of heritage conservation, and consideration of the need for greater protection of parks and gardens is in abeyance pending the HPR (see below).

In the meantime, planning authorities have used a range of available means to protect parks and gardens. Principally, that has been through the addition of a policy on registered sites in the local plan, in accordance with the lead given in PPG15 (see above). As Section 54a of the 1990 Planning Act had established the primacy of the development plan in determining planning applications, a policy included there for the protection of registered landscapes is potentially a powerful tool. However, its application depends on many factors.

The wording of individual local plan policies varies considerably despite the GHS issuing advice on wordings that have been approved by the Secretary of State.[19] The advice from the regional government offices and the decisions of local plan inspectors vary, as does the determination of local planning authorities to enforce the policy strictly. What constitutes harm is variously interpreted, with many applicants arguing that their development offers overall conservation gain despite clearly harmful elements. As is so often the case, the argument hinges on interpretations of balance. The problem for parks and gardens is still exacerbated by the confused planning status of the *Register* and the lack of experience among planners of this new heritage area.

In addition, planners have used other existing means to protect parks and gardens. Over 70% of registered sites are covered to some extent by Conservation Areas (CAs), although the argument persists that CA designation is not suitable for landscapes, despite being for areas of architectural and historic interest. Nevertheless the test for new development – that it should preserve or enhance the character or appearance of a Conservation Area – remains useful in rejecting harmful proposals, demolitions and tree-works, as well as for the opportunity it offers to suspend permitted development rights via Article 4 directions.

Listing of garden buildings is another helpful measure, but the number of examples on the Buildings at Risk registers indicates that it cannot easily enforce maintenance. In addition, thematically, Victorian park buildings are under-represented: far too many bandstands and shelters, glasshouses, drinking fountains, gates and railings remain unlisted and vulnerable.[20] In the case of a country house, the setting of a principal listed building can be argued to extend to the whole of the landscape designed to form its environment, and credence was given to this argument in PPG15.[21]

Scheduling, or local designation as an archaeological site, applies to a number of derelict gardens, but this can conflict with restoration when for example tree-planting or path-reinstatement proposals are put forward; archaeology has proved a good friend to parks and gardens, but the rigid protection it affords to fabric can prove problematic where there is still potential for garden conservation.

Habitat designations have been employed in some cases, and English Nature has taken an active interest since the mid-1990s in the habitat of parkland and wood pasture, but again the interests do not always match precisely. Scrub growth or dead trees will produce different management responses from ecologists and historic landscape managers. Where ecological interest derives from neglect, it can prove time-consuming, albeit essential, to agree a shared approach to conservation of the historic landscape.

In response to steady pressure about the lack of control over harmful development proposals within registered parks and gardens, a stop-gap measure was introduced in 1995 when the Department of the Environment made EH and the GHS statutory consultees on all planning applications which affect registered parks and gardens. EH is now consulted on Grade I and II* sites, while the GHS is consulted on all three grades. The GHS caseload remains the only indication of the level of planning activity affecting registered sites, and in the absence of statutory protection, the onus of preventing harmful development often falls on EH and the GHS, local amenity societies or planning officers to argue on a case-by-case basis.

While the planning status of parks and gardens has undoubtedly risen as a result of PPG15, local plan policies and statutory consultation, the *Register* remains a highly selective list. Even now that it has been upgraded and augmented, it contains only just over 1,500 sites. The buildings list contains nearly half a million, while county surveys of historic parks and gardens reflect a 1:10 ratio between registered gardens and those of county importance. This means that except where planning authorities have included county or local lists in the local plan, most historic parks and gardens in a given area remain largely unprotected. Furthermore, although county lists are being prepared by a number of planning authorities, the resources available for such non-statutory work tend to be sparse.

Poor-quality maintenance of landscapes can cause significant harm just as it can with buildings, and is probably impossible to legislate for. There is no equivalent in landscapes to the acute problems that can face a building, such as falling down, a point which is often misrepresented, and landscapes can survive poor maintenance in the form of benign neglect better than buildings. However, some of the most visible harm to historic landscapes is caused by neglect or by ill-informed change in management. A case in point is the adventitious growth of scrub and secondary woodland, or poorly located new tree-planting, which can obscure designed views (Figures 2 and 3). Technically, this is easy to reverse, but of course in public landscapes, felling of trees can be among the most contentious of repairs, and many public park restoration projects have had to carry out extensive public information campaigns in support of this element.

Parks and gardens, because outward views were so often an integral part of the design intention, are also vulnerable to development outside their boundaries. The Cadw *Register* for Wales usefully maps 'essential setting' and 'significant views', but this information is not included on the EH maps. As a result, the impact of development which affects a site visually but not physically can easily be overlooked (see below).

Grant aid

Grant aid for historic parks and gardens was first enabled as long ago as the 1953 Historic Buildings and Ancient Monuments Act, and the 1974 Town and Country Amenities Act extended the Historic Buildings Council's remit to gardens. However, no new money was made available to fund gardens, which limited the impact of the new powers. In 1977, the moribund Land Fund, set up by Hugh Dalton during the post-war Labour government, was revived and recast as the National Heritage Memorial

Figure 2 Peel Park, Salford, where 'Plant a Tree in 73' completely altered the character of the nineteenth-century park, blocking views and hurrying visitors along the paths and cycle-routes. (Parks Agency)

Figure 3 The designed view to Derryvore Church at Crom Castle on the shores of Lough Erne has largely disappeared due to natural scrub colonization along the edge of the lough. (The National Trust)

Fund, and this funded restoration projects at Painshill in Surrey, Biddulph Grange in Staffordshire and a number of other smaller projects. However to all intents and purposes it was the storms of 1987 that really increased the levels of funding.

With fallen trees littering the parks of London and blocking the capital's roads and pavements, the government quickly identified some £2 million to fund the clean-up. EH and the Countryside Commission established storm-damage repair programmes offering substantial grants, on the basis that replanting would not just reinstate fallen trees but also restore registered parks and gardens. For the first time, substantial grants of 75% were made towards the preparation of historic landscape surveys and restoration plans, ushering in the principle of a plan-led, whole-park approach. Restoration plans became restoration (or conservation) management plans because of the nature of landscapes: for example, except in the case of radical felling, the only way to achieve some elements of the restoration in many cases will be over a ten, twenty-five or even fifty-year cycle of tree management.

In 1990, further storms in the south and south-west resulted in both programmes being extended for a further period, eventually running until 1996. The amounts involved – the south-west region of the Countryside Commission, for example, spent around £250,000 annually for those six years – were small compared to present-day Lottery funds but were crucial in establishing a methodology for restoring historic landscapes. They were also important in introducing local authorities to the concept of conservation of historic parks and gardens, as they too benefited from the storm-damage repair grants.

EH followed up storm damage by piloting a garden grants programme with a budget of £200,000 per annum, while the Countryside Commission developed grants for the restoration of historic parkland under its Countryside Stewardship programme. However, it was the arrival of the HLF that transformed the conservation landscape (Figure 4).

Although the HLF had funded a number of garden projects for the National Trust and other charitable owners of country estates, the establishment of the Urban Parks Programme (UPP) in 1996 marked a fundamental change in approaches to parks and gardens. It based its work on the experience of EH and the Countryside Commission, offering grants for whole-park historic landscape surveys and emphasizing that restorations had to be plan led. But in addition, there were two far-reaching innovations. First, partly because there were so few public parks on national registers or inventories, it offered grants to sites irrespective of whether or not they were 'listed'. Heritage of local importance was assessed equally with national. Second, its aim was 'regeneration' of urban parks (the word was used in its first Annual Report in 1995), not merely repair. It was understood that, such was the condition of many public parks, merely repairing the damage to the fabric would not necessarily turn a site around, and repairs would be highly vulnerable to being damaged. What was needed to support restoration was often new infrastructure, and so the HLF put money into new toilets, new cafeterias, play areas, new feature gardens, as well as into the core heritage items such as bandstands, paths and railings. Not only did the UPP create a whole new heritage

Figure 4 The launch of the Heritage Lottery Fund's Urban Parks Programme in 1996 transformed conservation of historic parks and gardens.

sector, it changed ideas of what sites and what activities should be eligible for funding as heritage.

While the HLF had always eschewed policy making, in effect its grant-giving decisions and the requirements it has placed on applicants have had profound policy implications. The Secretary of State's directions in 1998 required that in distributing its funds the HLF ensure that projects addressed a much wider context than any previous heritage grant scheme: geographical equality, reducing economic and social deprivation, the needs of children and young people, sustainability. While some of these considerations were alien to traditional heritage grants, they were perfectly suited to the Urban Parks Programme, which rapidly became a favourite not only with the general public but also with politicians (Figure 5).[22]

It is arguable that funding public parks was a contributor to some of the major shifts in conservation policy that have resulted ad hoc from HLF's funding decisions. The emphasis on a plan-led approach, on public access and enjoyment, on contributing to economic and social regeneration, and a shift from an emphasis on fabric to wider considerations of public benefit, all are foreshadowed in the earliest grants to public parks.

Figure 5 Not just one but two Secretaries of State, for Culture and Health, attending the opening of the newly restored Coram's Field in 2000. (Land Use Consultants)

Restoration, conservation, re-creation and repair

In February 2006, English Heritage published a draft of *Conservation Principles*, which attempts to distil a unified set of rules across the whole of the conservation spectrum. In applying them to gardens, however, there are awkward areas where they do not quite fit. The planted fabric of gardens, as distinct from the built or constructed fabric of paths and buildings, cannot be accorded the same ascendancy as the fabric of historic buildings or archaeological remains. A mature tree, as much as a bedding display, is essentially ephemeral; it depends on replacement. In the case of landscape, conservation of fabric is a physical impossibility. This makes the distinction between 'maintenance' of fabric and 'restoration' or 'reconstruction' involving new introductions largely meaningless. The proper care of gardens is an unending and continuous process of adding to and removing fabric.

There are complications too even with the built fabric of a park or garden. Take the perennial 'bandstand question', regularly debated for the benefit of HLF trustees. Because of the years of dereliction and demolition out of which public parks are now emerging, the need to replace vanished architectural features is high on the shopping list of most applications. Many of these features were recorded in drawings and photographs; cast-iron structures were generally assembled from pre-cast kits from one of the major suppliers such as Macfarlane's in Glasgow, whose pattern books survive. Reinstating a cast-iron bandstand in its correct location, and in cast-iron from the original pattern, has been criticized as re-creation or even pastiche. The defence has been that the criticism fails to look at the park 'as a whole' as the heritage asset, rather than the individual component parts. When viewed that way, such interventions are no more than the equivalent of slotting a repair into a missing section of cornice on a historic façade, or patching a missing section of plasterwork in an interior.

It has also been important to recognize that new work can contribute to the conservation of a historic park, for example replacing a demolished Edwardian pavilion with a modern café in a contemporary style. In a heritage asset whose *raison d'être* is public enjoyment, 'regeneration' – encouraging the return of people – is a significant consideration, and repair alone will not deliver regeneration.

It is rare that conservation projects in parks or gardens return a whole site to a known earlier state: the sacrifice of mature trees alone would be too great. Other than in exceptional cases, such as Hampton Court Privy Garden (Figure 6), there is no equivalent to 'scraping back'. Because it is in the nature of landscapes to be in a process of continuous change and growth, and of design responses to those changes, it is generally inappropriate to think of such a return. In many cases, later planting or fabric has become too significant to destroy, although this does not mean that difficult decisions should be avoided (Figure 7). On the other hand, original fabric may not be relevant to the original design. For example, the

Figure 6 A rare example of 'pure' restoration – scraping back years of change at Hampton Court in order to restore the formal layout of the Privy Garden designed for William and Mary. (Parks Agency)

Figure 7 One of the eighteenth-century Cedars of Lebanon at Painshill – impressive in its own right but now dwarfing the designed landscape. (Painshill Park Trust)

Figure 8 The impact of tree removal at Ashton Court, Bristol: decisions could only be agreed after lengthy public consultation. The illustration on the left shows the view of the west lawn before the removal of the large maple tree, with the view after tree removal on the right.

famous Lebanon Cedars at Painshill, venerated as veterans from the original eighteenth-century design, are highly respected but it is arguable that they now detract from the eighteenth-century design, because they have outgrown the landscape and dwarf the features of which they once formed the setting (Figure 8).

Trends in casework

It would be good to say that, with PPG15 (1994) and the introduction of local plan policies on parks and gardens, with statutory consultation (1995), and with the growing government awareness of the importance generally of greenspace,[23] major threats to historic parks and gardens are a thing of the past.

It is true that compared to the late 1980s, when a combination of a property boom and a brand new *Register* with which planners were still unfamiliar, resulted in a rash of utterly misconceived and damaging proposals for golf courses, country clubs and hotels, there is now a good deal more circumspection on the part of developers. Croome Court in Worcestershire for example, now being restored by the National Trust and HLF, nearly succumbed to a woeful golf-course scheme which would have packed 27 holes side by side into the Grade I parkland, and a number of sites were left scarred by intrusive golf-course developments (Figure 9). A few high-profile appeal decisions helped to establish a kind of case-law on what was and was not appropriate, what was and was not 'restoration' and what was and was not a proper approach to introducing change in a historic park or garden (Figure 10).[24] In the case of golf, reports by EH, the Countryside Commission, GHS and Georgian Group, and a number of local pieces of supplementary planning guidance, helped to stem what had been an uncontrolled flood of planning applications.

At the same time, however, the indeterminate status of parks and gardens remains a source of uncertainty on the part of owners and developers. The *Register* is still regularly referred to mistakenly as non-statutory (it was enabled by statute, but there are no additional statutory powers or requirements associated with it), and as a result inevitably it is seen as being a lesser material consideration than others. The status of 'should' in PPG15, as in 'local authorities should protect historic parks', remains maddeningly ambiguous.

Perhaps the most pernicious threat to landscapes associated with historic buildings came with the concept of 'enabling development', which grew up in the late 1990s. This almost invariably led to a stand-off between the importance of a building requiring repairs and the available space in which to construct development to fund those repairs. As the development space tended to be in land closely associated with the main building, this meant

Figure 9 Orchardleigh in Somerset, where a golf-course development, shown here during construction in 1990, has had significant impact on the form of the designed landscape. (Garden History Society)

Figure 10 The Grade I park at Warwick Castle, threatened by golf and hotel development in the late 1980s, was saved in a landmark appeal decision which hinged on a critical analysis of the conservation gains offered by the developer. (Garden History Society)

in effect the historic park or garden. In such cases, the low status of the *Register* compared to that of the buildings list proved a fatal flaw. Walled gardens were particularly vulnerable, but the main issue was development designed to have a view of the main house. Worse still was the threat to garden land that was not registered. The infamous case of Downe Hall in

Figure 11 Enabling development at Downe Hall, on the edge of Bridport: the landscape was unregistered when the application was first submitted, and the application proved impossible to halt after attention was finally drawn to the importance of the landscape. (Garden History Society)

Bridport illustrates the fate of an unregistered garden of a listed building. Although it was subsequently added in recognition of its historic interest, the landscape's fine quality was given short shrift when the case was made for its development to enable the repair of the roof of a listed building. Despite representations locally and nationally, no one picked up the importance of the landscape until far too late (Figure 11).

EH's guidance on enabling development, with its strict tests, has to some extent helped to stem the flow of these often very damaging applications.[25] However, despite that excellent guidance, many of the threats posed by enabling development have not gone away. The GHS still sees many applications for development within walled gardens, for example, which are effectively enabling development, although there is now a marked reluctance on the part of applicants, and in some cases planning authorities, to acknowledge them as such. In most cases, at the root of the problem lies a tendency on the part of some local authority conservation officers – and even some EH Inspectors – to place greater weight on the perceived 'needs' of a listed building than on the value and historic interest of the surrounding designed landscape. This in turn goes back to the relatively low status of the landscape designation. Despite all the welcome talk about a more 'holistic' approach to the historic environment since the publication of *Power of Place*,[26] it appears that in practice many minds remain firmly compartmentalized. Hence, for example, the recent application for housing in the walled garden forming an integral part of the

Grade I eighteenth-century landscape at Croome Court, which is justified on the basis of the need to repair a semi-derelict wing of the adjacent mansion. The planting records, and the involvement of Capability Brown in its design, make the walled garden at Croome one of the most important in the country; yet the inability of the system to recognize the importance of a 'space', *per se*, as a historic asset, remains highly problematic.

While many of the casework 'themes' remain depressingly familiar from ten or twenty years ago, there are also new threats. Concerns about climate change and energy security have inevitably led to proposals for various types of sustainable generation projects, which can range from large-scale wind farms, to solar panels on buildings within the designed landscape, or hydro electric schemes affecting picturesque designed landscapes in Scotland and other upland areas. Some may have relatively little impact, but others, such as wind farms, even when some distance from a registered landscape, can have a significant adverse impact on views and settings of designed landscapes. Despite some encouraging decisions by the Planning Inspectorate, it is far from clear what the eventual impact of these schemes will be on designed landscapes. Similar considerations apply to the steady stream of consultations received by the GHS on applications for telecommunication masts and aerials. In some instances, they are relatively well designed and thought has obviously been given to the potential visual impact on registered landscapes or their settings; but in other cases there is a startling lack of awareness – for example the proposal to construct a lattice tower on one side of William Kent's great vista from the landscape at Rousham, Oxfordshire (Grade I) to the eye-catcher beyond the Cherwell; or a proposal for a mast on the high ground immediately adjoining the registered site boundary of Compton Verney, Warwickshire (Grade II*), on which the GHS and EH were not initially consulted because the planning authority did not consider the mast to impact on the setting of the landscape.

Historic landscapes associated with buildings in institutional use continue to give rise to development proposals which can have significant adverse impacts on the landscape. In the past, pressure has usually been associated with the need to provide additional facilities, especially where the institution is a school. Increased competition between such establishments has not led to a diminution of such applications, but equally there have been a number of cases where institutions have, for various reasons, moved out of historic buildings. These situations have been handled with varying degrees of success, but in some instances, such as Rousdon in Devon (Grade II), where a good development brief was produced, the school buildings have been removed and new residential development has been designed in such a way as to have a reduced impact on the designed landscape and setting of the listed buildings. More problematic, perhaps, are the sites which were designed as institutions, such as asylums or hospitals like Severalls at Colchester, Essex, which have been closed and declared redundant. In many instances, the owner or developer has no desire to retain the unlisted elements of the original buildings, and frequently little weight is given to the landscape, which was usually an integral part of the overall design, with a therapeutic role of its own. The temptation to see such sites as 'brownfield' sites ripe for intensive redevelopment is clearly great, but as a result we are in serious danger of depleting our heritage of this type of landscape.[27]

Heritage Protection Review

There has been a tendency over the past two years to see the Heritage Protection Review (HPR) as something EH has undertaken of its own volition, rather than as a political exercise driven by government. Despite the impressive performance of the Department of Culture, Media and Sport (DCMS) Minister, David Lammy, at a joint DCMS/Office of the Deputy Prime Minister (ODPM) presentation on the HPR last autumn, there seems to remain an obstinate misunderstanding of heritage and its social and economic contributions. In part this can be attributed to the division of departmental responsibilities and lack of 'joined-up thinking' between DCMS and ODPM, but in the end it is hard to escape the impression that heritage poses fundamental ideological problems for this government. Despite all the evidence produced to the contrary, it persists in giving the impression that heritage concerns are a brake on economic growth and regeneration. The huge benefits of heritage-based tourism are routinely ignored, and the mantras of 'simplicity' and 'transparency' are used as an excuse to jettison years of experience and good practice. If not exactly guilty of an Orwellian desire to control history, 'New Labour' has always distanced itself from the old.[28] Against this background, it is not surprising that the HPR gives considerable grounds for unease.

For historic designed landscapes, there are aspects of the HPR which are welcome – in particular, the understanding that the value and significance of the historic environment lies in the whole, rather than in unconnected component parts. This is a welcome philosophical change, and the concept of a new, unified National Register to subsume the existing designation regimes of scheduling, listing and registering seems a logical corollary of this new understanding. However, it looks ominously as if the government has not sufficient courage to carry the logic forward to the control regimes relating to the existing three 'levels' of designation and establish aspects of heritage identified as being of national significance a similar degree of protection. Instead, the existing anomalies and inconsistencies appear likely to be carried forward into the new system, with the result that designed landscapes will still be under-protected relative to other areas of the historic environment.

It is hard to understand how the government considers this unsatisfactory hybrid system to be any more fit for purpose than the existing designation system, or why it should suppose that the control regime will be any more transparent or intelligible to owners and planners. We are now promised a system of colour-coded maps – green for a statutory material consideration, red for archaeological and monumental designations with the highest level of control, blue for controls linked to structures which are neither monuments nor archaeology, and yellow for a proposed new level of control indicating 'flexible management' of certain landscape or archaeological features. There seems to be enormous potential for confusion and conflict with other, predominantly natural environment designations, which also use a similar colour-coded map base.

The government's lack of understanding of the fundamentals of heritage designation and protection is clearly illustrated in its attitude to the grading of items to be included on the National Register. At present, listed buildings and registered parks and gardens are categorized into one of three Grades, I, II* or II according to their level of special historic significance; scheduled monuments are not graded. Ministers consider this system to be confusing to the public, and insist that the new system will instead have just two Grades, 1 and 2, which will be applied to all heritage assets. This has been a consistent feature of the DCMS proposals, and despite representations from, among others, all the national amenity societies, there has been a remarkable unwillingness to respond to reasoned criticism based on many years' experience.

Under the government's proposals, items presently graded at I and II* will be regraded at Grade 1, while items presently graded at II will become Grade 2. The conflation of Grades I and II* makes little sense for listed buildings, but when the same process is applied to the more selective (because of the smaller number) designation of historic designed landscapes, it will result in a devastating devaluation of the 'gold standard' of Grade I, that is to say landscapes considered by EH to be of special historic interest not only in the national context, but in the international context. Landscapes of international importance, such as Stowe or Blenheim (the latter also being designated a World Heritage Site), will be placed in the same grading band as sites of considerably less historic interest, including those which EH inspectors felt to have been 'borderline' between Grade II and Grade II*, but which on balance just made the higher grade. When considering grading as part of the Register Upgrade Programme, inspectors did not consider the possibility of future grade conflation, with the result that for consistency, the grading exercise should be undertaken afresh.

The problem perceived by ministers – for which there appears to be little concrete evidence – would seem to be better addressed through a programme of public education, rather than the introduction of an unnecessarily restricted grading system which cannot hope adequately to reflect the different levels of historic interest attaching to different designed landscapes.

Where the new designation system does promise to be an improvement on the existing situation is in the way that the descriptions of the various 'assets' are written – and for this, the sector should undoubtedly thank EH. Interestingly, the new descriptions will be closely modelled on the form developed for the second edition of the *Register of Parks and Gardens*, which begin with a succinct 'statement of significance' explaining what makes this site of special historic interest, before moving on to an outline of the historic development of the site and a more detailed description of the landscape as it survives today. This level of information for listed buildings and scheduled monuments would be of great benefit to owners and planners, and would undoubtedly help in encouraging greater understanding of the properties in their care. But with some 500,000 existing listings, the process of revision and rewriting can only realistically be considered an aspiration – unless DCMS is prepared to show its commitment to heritage protection by securing appropriate levels of funding from the Treasury.

Education and training

As is increasingly being recognized by funding bodies, one of the keys to sustainable management of the heritage is education and training. The craft skills of gardening and horticulture have been promoted by English Heritage, GARLAND (the Garden and Landscape Heritage Trust for the Advancement of Education and Training) and the professional bodies, and grant-aided by the Heritage Lottery Fund, not only via grants for specialist staff posts but also through a recent garden bursary scheme. In January 2006, HLF announced a grant of £721,000 to fund a partnership of seventeen heritage organizations led by English Heritage to run a 'Historic and Botanic Gardens Bursary Scheme'. This will offer placements to help develop horticultural skills for historic parks, gardens, designed landscapes and their plant collections. It has not come a moment too soon, as the sector has been struggling with recruitment, with wages for craft-gardeners still shamefully low.

However, horticulture aside, there remains a serious lack of educational opportunities for planning and conservation professionals, estate managers and private owners. It is a sad fact that experts on historic buildings often have very little understanding of the nature and complexity of the designed landscape settings of those buildings. A snapshot of the past 25 years would have to record the rise of a number of excellent courses on historic landscape conservation – at the Architectural Association, De Montfort University and the University of York – and their demise in the last five years. There are a number of garden history courses – Birkbeck College, London offers a diploma/certificate and a post-graduate diploma convertible to an MA in Garden History, while a commendable Garden History MA has been established at the University of Bristol – but these courses are predominantly art-historical, rather than conservation or management-based. The University of Bath is considering setting up a more conservation-oriented course in the near future, but the fact remains that there is now less opportunity for training in the understanding and conservation of historic landscapes than there was ten years ago.

Conclusion

Despite major advances during the 1990s, which have seen historic designed landscapes enjoying unprecedented recognition as part of the national and local heritage, misunderstanding still plagues their conservation. Their repair and maintenance, and their protection from harmful development, are both erratic. Despite the injection of grant aid for capital works, many public parks remain under continuous threat of reductions in revenue spending from local authorities. In the absence of expertise within local government, and of resources in expert bodies, consents for damaging development will inevitably continue to be granted; and in the absence of equal protection in law, historic parks and gardens will inevitably continue to suffer in the balance against other types of heritage asset.

Biography

David Lambert
David Lambert is a director of the Parks Agency, and was previously conservation officer for the Garden History Society. He has been closely involved in the Heritage Lottery Fund's public parks grant programme since its inception. He has served as a special adviser to several parliamentary inquiries, including that on public parks. He is a member of the advisory panels on parks and gardens for both English Heritage and the National Trust.

Jonathan Lovie
Jonathan Lovie holds degrees from the University of St Andrews, including an MPhil awarded for a thesis based on original historical research. Since 1994, he has developed a consultancy undertaking historic landscape research. Between 1998 and 2004, Jonathan was employed by English Heritage as a Consultant Register Inspector, with responsibility for upgrading the *Register of Parks and Gardens of Special Historic*

Interest in the South West of England. Since 2004 he has been Principal Conservation Officer and Policy Advisor for the Garden History Society. He lectures on research and conservation of historic landscapes at the Architectural Association, Bath University and Oxford University.

Notes

1 Hussey, C., *The Picturesque: Studies in a Point of View* (1927).
2 Clark, H. F., *The English Landscape Garden*, Pleiades Books, London (1948).
3 Stroud, D., *Capability Brown*, Faber & Faber, London (1950).
4 Hadfield, M., *A History of British Gardening* (1960). The first edition appeared under the title *Gardening in Britain*.
5 Hunt, P. (ed.), *Shell Book of Gardens*, Phoenix House, London (1964).
6 Hussey, C., *English Gardens and Landscapes 1700–1750*, Country Life Ltd, London (1967).
7 Strong, R., *The Renaissance Garden in England*, Thames & Hudson, London, (1979).
8 Jacques, D., *Georgian Gardens: The Reign of Nature*, Batsford, London (1983).
9 Elliott, B., *Victorian Gardens*, Batsford, London (1986).
10 Conway, H., *People's Parks*, Oxford University Press, Oxford (1991).
11 Williamson, T., *Polite Landscapes: Gardens and Society in Eighteenth-Century England*, Sutton, London (1995).
12 For a short account of the rise of historic landscape conservation since the nineteenth century, see Batey, M., Lambert, D., and Wilkie, K., *Indignation: The Campaign for Conservation*, Thames Landscape Strategy, London (2000).
13 Historic Buildings Council Annual Report, 1973.
14 Department of the Environment/Department of National Heritage, *Planning Policy Guidance Note 15, Planning and the Historic Environment* (PPG15), HMSO, London (1994), para. 2.24.
15 *Power of Place: The Future of the Historic Environment*, English Heritage, London (2000).
16 English Heritage, *Sustaining the Historic Environment*, English Heritage, London, (1997).
17 London Assembly Environment Committee's report, 'Crazy Paving' (September 2005).
18 Pendlebury, J., 'Historic Parks and Gardens and Statutory Protection', Department of Town and Country Planning, University of Newcastle upon Tune, Working Paper No. 44 (1996).
19 Garden History Society, *Advice on the Protection of Historic Parks and Gardens in Development Plans* (first published 1992, latest edition 2002).
20 See *GreenSpace*, Public Parks Assessment (2000).
21 PPG15, paras. 2.16–17.
22 Lambert, D., 'The Heritage Lottery Fund's Urban Parks Programme', *Transactions of the Ancient Monuments Society*, Vol. 46, pp. 83–96 (2002).
23 For example: House of Commons Environment, Transport and Regional Affairs Select Committee, *Town and Country Parks* (1999); DETR, *Our Towns and Cities: the future* (the Urban White Paper) (2000), pp. 74–6; DTLR, *Green Spaces, Better Places*: the final report of the Urban Green Spaces Taskforce, Department for Transport, Local Government and the Regions, London (2002); ODPM, *Living Spaces: Cleaner, Safer, Greener*, HMSO, London (2002).

24 Lambert, D. and Shacklock, V., 'Historic Parks and Gardens: a review of legislation, policy guidance and significant court and appeal decisions', *Journal of Planning and Environment Law,* July 1995, pp. 563–73.

25 *Enabling Development and the Conservation of Heritage Assets: Policy Statement: A Practical Guide to Assessment,* English Heritage, London (2001).

26 *Power of Place: The Future of the Historic Environment,* English Heritage, London (2000).

27 See Garden History Society, publication ongoing, *Planning Appeals Digest* for a summary of appeal decisions affecting historic parks and gardens.

28 For example, *Green Spaces, Better Places,* the report of the DTLR's Urban Green Spaces Taskforce (2002), was remarkable in avoiding the word 'historic' in dealing with urban parks, or acknowledging that people liked existing parks for their mature planting and historic buildings. Memorably, it opined that there was 'too much "old hat"' in Victorian parks' (p. 18).

SAVE Britain's Heritage and the Amenity Societies

Adam Wilkinson

Abstract

Effective conservation depends upon a committed, educated and inspired voluntary sector, capable of handling casework on a daily basis, reacting to urgent threats and undertaking or commissioning research projects as circumstances demand. SAVE Britain's Heritage is part of this movement; yet it is distinct in its methods. It is long established but unconventional, playing its role in a bold fashion now recognized as the hallmark of its operations. Since SAVE's founding in 1975, threats to the historic environment have become ever greater and more sophisticated, prompting SAVE (and the amenity movement as a whole) to become ever more professional and effective with their resources. As new threats emerge, SAVE and other amenity bodies incur greater workloads in responding to development proposals and in educating and informing decision-makers and the public. This paper first considers the value and role of amenity societies generally before exploring the work of SAVE Britain's Heritage, and the manner in which it applies its ideas, convictions and resources.

Introduction

The 'amenity societies', as we have come to know them, exist to educate, inform and campaign as they find necessary in promotion or defence of the nation's architectural and related heritage. They are numerous, but generally well focused and always heavily dependent upon volunteer support. Many other countries have their conservation organizations but several aspects of the British scene merit particular mention. Here, in contrast to our European neighbours, the societies have generally emerged to represent particular periods of construction, or specific forms of artistic or architectural expression. They range from tiny, informal and focused bodies to organizations with constitutions crafted to allow and encourage structured debate within the membership, and to report this widely within the conservation sector and beyond. Thus, for example, we have the Victorian Society and the Georgian Group dealing with broad periods, each containing numerous architectural styles, with both groups having an interest in most towns and nearly all cities in the UK. At the other end of the spectrum lie the highly focused bodies, no less valuable in their fields, such as the Folly Fellowship, the Mausolea and Monuments Trust, the Tiles and Architectural Ceramics Society and many more.

Over time they have become so valuable a repository of informed opinion that successive governments have drawn certain of them ever closer. Five longer-established societies have, since 1972, enjoyed a statutory right of consultation on all applications involving demolition or part-demolition of listed buildings. A sixth, added by the Deputy Prime Minister in October 2005, recognizes the importance of buildings of the twentieth century and has a highly informed understanding of the Modern Movement. The six, in order of their dates of founding, are:

Society for the Protection of Ancient Buildings (1877)
Ancient Monuments Society (1924)
Georgian Group (1937)
Council for British Archaeology (1944)
Victorian Society (1958)
Twentieth Century Society (1980)

SAVE Britain's Heritage is similar to many other amenity societies in the sense that it depends for much of its impact on volunteer effort, but it has no boundaries of period or building type. SAVE was founded in 1975 by a group of planners, architects and journalists in reaction to the destruction of historic buildings, with the clear sense that their concern was shared by the wider public, and that the press could be encouraged to articulate this concern. The year 1974 witnessed *The Destruction of the Country House* exhibition at the Victoria and Albert Museum, curated by John Harris and Marcus Binney. A press release was sent to every local paper covering lost country houses in their respective areas, and a remarkable number of local newspapers followed up with a story.[1]

Although the press release remains an important tool in SAVE's armoury, the charity has a free-ranging role, its primary work being to campaign for threatened individual historic buildings or entire categories considered under threat. Free of the burden of statutory casework falling upon the six societies listed above (though lacking the state income this provides), SAVE is able to act where it feels most benefit will result.

All income comes from donations and the sale of publications, allowing the organization a free hand in what it says, when, and how. Its trustee body and advisory committee comprise a potent combination of journalists, architects, architectural historians, planners and conservationists. Both these governing groups allow the employed staff a relatively free hand, making rapid action possible, although their constituent members make themselves available to staff for direct and informal discussion of cases in their own areas of expertise. The advisory committee plays an essential role in helping decide which cases to focus upon – since the organization receives many more requests for help than it can possibly respond to.[2]

Aside from the press release, the other main weapons in the SAVE armoury are the 'lightning report' (documents on threatened buildings), the thematic report (looking at focused issues and building types – most recently historic law courts), exhibitions to profile issues, and a twice-yearly newsletter in whose pages polemic rules. Behind all of these is the positive ethos of showing what can be done to reuse threatened buildings. These tools have been highly successful over the past 30 years in highlighting threats to a range of historic buildings, from Scottish castles to public lavatories.

The way heritage is valued has changed radically since 1975 – Victorian buildings are no longer despised, buildings of the Modern Movement are better understood, and SAVE's own views on some of the buildings it originally campaigned against – such as the Gateshead car park – have been reversed.[3] However, it is a common misconception that we live in an enlightened age and that the major battles have been won. The everyday experience of SAVE is that present threats are often more complex than the fight for any single building.[4]

Mentmore and the National Heritage Memorial Fund

In 1977, SAVE launched a typically vigorous campaign to prevent the break-up of the great collection at Mentmore, one of the Rothschild prodigy houses, the design of which was based on Wollaton, as interpreted by Joseph Paxton and built between 1852–54. The campaign ended in disaster as the government refused to finance the purchase, only to then spend as much on individual items from the collection for museums as it would have cost to buy both the house and its collections.

The National Land Fund was set up in 1946 by Hugh Dalton, the then Chancellor of the Exchequer, funded by the sale of massive quantities of war stores, with £50 million set aside for the preservation of famous historical houses and stretches of countryside. Mentmore was debated in Parliament, but the contents of the house were auctioned at Sotheby's and this individual battle lost. However, Marcus Binney, founder and president of SAVE, gave evidence to the Parliamentary Expenditure Committee, suggesting that the National Land Fund should be reconstituted under independent trustees. When the committee concurred, a train of actions was put in motion, leading to the creation of the National Heritage Memorial Fund, now the parent body of the Heritage Lottery Fund.

Parliament remains an effective means of raising awareness of issues and forcing the government to consider them, the most recent example being the Culture, Media and Sport Committee's damning indictment of the government's stewardship of heritage in recent years[5] – SAVE and the amenity societies gave evidence to the committee, and SAVE has contributed to other parliamentary inquiries.

Barlaston Hall – theory into practice

The Barlaston case is a powerful indicator of what can be achieved by the voluntary sector on the scarcest of resources in the face of opposition from established interests. Barlaston, a handsome Palladian villa, the work of Sir Robert Taylor (1714–88), sits on a prominent site in north Staffordshire. Owners Wedgwood had applied to demolish the building, which was in disrepair and facing potential subsidence through planned coal mining. The case went to a public inquiry. SAVE and the Georgian Group were disparagingly referred to by Wedgwood's QC as the 'United Aesthetes' before he challenged SAVE to purchase Barlaston for £1. After a transatlantic phone call by SAVE's then Secretary Sophie Andreae (frantically pumping coins into a payphone, this was 1981) to Marcus Binney, it was decided upon. The offer was accepted and the Inspector sportingly offered the ten pence deposit, perhaps pleased he would not have to write a lengthy inquiry report.

The Historic Buildings Council offered a grant towards emergency work, mainly to get a roof on the building, and then offered a further £150,000 towards the full repair. However, this payment was conditional upon the National Coal Board making a contribution towards preventative measures to protect the building from further subsidence. The Board did not initially make any contribution, and it was not until legal proceedings were initiated by SAVE that the matter was resolved.

Royal Aircraft Establishment at Farnborough

The Royal Aircraft Establishment at Farnborough is the cradle of British aviation, and hosts possibly the finest collection of flight research and development material in the world. That the site was not widely known about is testament to the effectiveness of military secrecy and played directly into the hands of the Ministry of Defence, which wished to sell to the highest bidder. The Ministry of Defence (MoD) employed consultants to argue successfully against the listing of many buildings on the site, and

Figure 1 Barlaston Hall, Staffordshire, taken on and repaired by SAVE in the face of obstructive officials, a massive shortage of cash and sinking ground.

found itself unable to work creatively with the Farnborough Air Sciences Trust (FAST), which had been established to preserve the collective memory of the establishment and to show how the site and its buildings might be reused.[6] The site was purchased by Slough Estates, who rapidly fenced off the central area containing the majority of the listed buildings. Everything else, including buildings that should have been listed, was demolished to make way for a new business park. At first, Slough's plans were to retain only two listed buildings on the site: Q121, a 1935 wind tunnel with a vast and beautiful return air duct and a testing space large enough to accommodate a full-scale aircraft; and R133, the 1939 transonic wind tunnel, for testing the behaviour of scale models close to the speed of sound. Both these remarkable buildings, and others, retained their original machinery, unlike other known historic aviation sites here or abroad.

SAVE's approach was to secure protection through listing, and then work upon viable alternative uses. To this end, a 'lightning report' putting the site in its international context was produced and launched to the press and all efforts were made by FAST and SAVE to upgrade listing grades.[7] After much pressure, both Q121 and R133 were listed at Grade I and the earliest surviving wind tunnel building R52 at Grade II. To encourage

Figure 2 The 1930s 24 ft wind tunnel at the Royal Aircraft Establishment
Farnborough is a monument to the age of powered flight and an engineering marvel,
built to tolerances that can only now be dreamed of. It is capable of testing a full-
scale fighter aircraft at speeds of up to 120mph, and is now listed at Grade I. It
retains all of its original machinery. (FAST)

protection of unlisted buildings in the central area, SAVE produced a
second report (with English Heritage grant support) showing how they
could be profitably reused. Today, almost all buildings of merit in the
central area have been retained and are being restored, with the developers'
marketing suite occupying the former weapons test building.

Current threats

Threats from government

At present, government appears to lack real interest in heritage. Neither Tony Blair's recent letter to Culture Minister Tessa Jowell on her departmental priorities, nor her response, mentions it.[8] Government policy calls for engagement with local public opinion and all sections of society, but seems to fail to recognize that the amenity groups are the expression of informed public opinion, along with the many local history, civic and related societies evident in most towns and cities.

Heritage seems to languish without an effective champion in government. Many departments and state bodies appear to function in a heritage vacuum, including the National Health Service (NHS). There is a manifest lack of co-ordination in policy making – for example the use of the Private Finance Initiative for building procurement is inherently biased against smaller, more complex historic buildings, be they hospitals, schools or law courts.[9] In the rushed divestment of historic buildings from the public estate, conservation appears to receive the barest attention. Government housing policies threaten swathes of Victorian housing and their communities through the Pathfinder initiative, and if government does not take heritage seriously it can hardly expect the private sector to do so.

Threats from within

A constant challenge for the amenity societies is that of how to be a critical friend to bodies such as English Heritage and the Heritage Lottery Fund. The former has suffered in recent years through declining funds and the obligation to become 'customer focused'.[10] The cut in the level of government funding available to the organization cannot be without negative effects for heritage as a whole – many in the sector feel that English Heritage has been lost as the sector's champion and leader.[11] There is an urgent need for English Heritage to be re-energized and re-invigorated, focusing on the core work it can do so well. However, while it is in a delicate position it is difficult for its friends to comment on where its weaknesses lie without its opponents jumping on these points.

The Heritage Lottery Fund (HLF) has undoubtedly been a leader in the sector in terms of both funding and setting standards – with its massive funding, it has the ability to force the sector to act as it sees fit. Not all of this has been as positive as one might have hoped. SAVE holds that one unintended result has been the emergence of an army of consultants to help ensure that all the boxes are ticked. Consultation costs can be prohibitive for smaller voluntary organizations taking on large rescue projects – SAVE is aware of one case where £200,000 was spent on consultants for a £2 million project before any funds were applied to the building.[12] While there is a need to properly understand a building before starting work on it, the culture of paperwork appears to have gone so far as to make the HLF effectively inaccessible to smaller organizations.

Strong educational demands are placed on any project by the HLF. This frequently causes a relatively straightforward repair project to become buried in supplementary projects largely irrelevant to the restoration, thereby draining time and funds, particularly where the project is sponsored by a small organization. While it is desirable that the public should have access to historic buildings and learn about them, it seems absurd to refuse support to repair a building because the educational or access package that goes with it is judged weak. Frustratingly, applicant organizations are reluctant to speak out, fearing loss of funding.

At the level of local government, SAVE remains concerned at the number of buildings at risk in local authority ownership. Local authorities should be leading the way in dealing with both these buildings as well as privately owned buildings at risk: they have the powers under the Planning (Listed Buildings and Conservation Areas) Act of 1990, but lack the confidence to make use of them. Recently, in the Newsham Park area of Liverpool, the

Figure 3 One of the houses on Prescot Drive, Liverpool, by Newsham Park, owned by Liverpool City Council. LCC refused to sell off this and other buildings in spite of handsome offers from residential developers to take them on and repair them.

local authority was forced to consider selling historic buildings in its possession that had fallen into disrepair, through the novel use of a Public Request to Order Disposal on the advice of the Empty Homes Agency. Under the 1980 Local Government, Planning and Land Act, publicly owned land that is harming the amenity of an area can be forced onto the market.

It is still the case that in certain local authorities conservation is seen as a luxury. Conservation officers' advice is rejected on the weakest of reasoning. The salaries offered to conservation officers are seldom good, but SAVE is gravely concerned to find evidence of salaries so low as to suggest the role is considered marginal or irrelevant. Conservation officer roles are known in some case to go to trainees and assistants who lack the education, training or enthusiasm required to fulfil the role. Bob Kindred's excellent study of salaries in the sector backs up this anecdotal evidence.[13] This is a specialist field, requiring trained specialists.

Five current issues

Listing

The current listing regime for the statutory protection of listed buildings is in SAVE's view failing many deserving buildings. At present, the only way a building can be listed is through the *ad hoc* process of spot-listing. Gone are the area studies and even the more recently instituted thematic studies, which, although seen by some as flawed, at least lent a degree of intellectual rigour to the process.

SAVE is compiling a dossier of buildings rejected for listing, in an attempt to see if any themes emerge. So far, examples range from the absurd to the worrying. An example of the absurd is a refusal to list on grounds of radical change, when it appears recent photographs were being compared to a historic picture – of a different building. A worrying example is that of a Georgian terrace being turned down in spite of the fact that neighbouring buildings of lesser interest are listed.

SAVE wonders whether at the root of this might be pressure to limit the number of listed buildings. The notification to a building's owners of requests to spot-list can be the death sentence for the building – SAVE is aware of cases where owners have deliberately vandalized or demolished their buildings when notified of the possibility of spot-listing – intended to prevent precisely that. A particularly galling example is the Rope Works in Bow, a handsome collection of Victorian industrial buildings vandalized to make listing impossible the moment a date had been agreed for an inspection.[14]

The forthcoming White Paper represents a once-in-a-generation opportunity to reform the listing system. Greater transparency in the listing

Figure 4 The General Market building at Smithfield has been granted a certificate of immunity from listing in spite of its obvious architectural and historic interest.

Figure 5 The ropery in Bow was an important part of the military industrial complex in the nineteenth century. When faced with a pre-arranged spot-listing inspection, the owner moved to ensure it was rendered unlistable through partial demolition.

and delisting process is to be welcomed, but SAVE remains to be convinced that the creation of a unified register and the amalgamation of Grade I and Grade II* listed buildings into a single grade will make the system easier to understand and help it operate any better.

Conservation areas and demolition control

Conservation areas remain the only area-based form of controlling development in historic areas. However, their effectiveness has been severely weakened over the years by many factors, including: the Shimizu decision; the failure of local authorities to take enforcement action; a disregard for the rules by certain members of the developer community; the status of conservation officers in local authorities; and the precedence given to economic development over conservation by members within certain local authorities seemingly incapable of seeing the two working hand-in-hand. The Shimizu decision, in effect, discounted alterations and partial demolition from being classified as demolition, thereby not requiring conservation area consent. Government has consistently failed to live up to its promises to correct this aberration.[15] In this way, the character of many of our historic town centres and fine Georgian and Victorian suburbs is being eroded. SAVE receives a constant stream of calls from individuals and organizations for assistance in the fight to retain the character and sense of place that makes where they live special. Although cases within conservation areas can be hard to fight, they are often the most rewarding to win, giving both communities and local authorities more confidence for future battles.

A recent example is 119 Poplar High Street in east London. The developer argued strongly for demolition, on the grounds that this typical 1880s corner building did not contribute to the character and appearance of the area, as there were better buildings in the area and it was structurally unsound. The local authority officers agreed with this, but its members refused the application. The developer appealed. At the ensuing public inquiry, the local campaigners put up a fine resistance, using evidence from SAVE and the Victorian Society as evidence of the building's positive contribution and its potential for economic reuse. The entire neighbouring terrace signed a petition in support of the building, and sensibly the Inspector rejected the appeals. This is a rare case – frequently such applications are steamrollered through in the name of progress.

Demolition of any unlisted building, or one not in a conservation area, is not considered 'development'. The owner simply serves notice on the local authority of the intention to demolish. This means that there is little or no accounting for a building's local significance and contribution to a place's uniqueness, and in some cases, for local listing. Consequently, a great many decent and well-loved historic buildings, perfectly capable of

Figure 6 119 Poplar High Street, precisely the sort of building conservation areas are meant to protect, saved from demolition by local grass-roots action – and a bit of help from SAVE and the Victorian Society.

Sketch of repaired facade for Dalston Theatre and 8-10 Dalston Lane.

Figure 7 Unlisted and not in a conservation area, these late, now ruinous Georgian houses and Victorian theatre on the Dalston Lane, London face demolition, in spite of their clear potential for repair and reuse.

economic reuse, are demolished. The St William of York deposition, a campaign led by Moyra McGhie, represents a grass-roots movement to change this situation by attempting to have demolition classified as a form of development under the General Permitted Development Orders. This would ensure at least a degree of discussion before a building is lost.

Buildings-at-risk

Buildings-at-risk remain a huge problem. SAVE's own register of Grade II listed buildings-at-risk only covers a small proportion of them, and yet has around 800 entries (unlike the English Heritage Register of Grade I and II* buildings, it is aimed directly at people seeking a repair project, so it does not cover structures that have no possible new use). Every year, more are added as our information improves.

Local authorities are not actively encouraged to use the powers granted to them by Parliament to deal with listed buildings-at-risk in their areas. These include both those in the Planning (Listed Buildings and Conservation Areas) Act 1990, Sections 54 and 48, the Town and Country Planning Act 1990, Section 215, as well as some of the more obscure powers under the Housing Acts. Section 215 notices are especially useful as they do not incur any liability on the part of the local authority (the powers under Section 215 concern tidy sites and can be applied to buildings that are showing the external signs of deterioration).[16]

The obstacles to the use of these powers come in three forms: firstly, from authority members, who are either unaware of the issue or afraid that the use of what can end up being quite draconian powers in an area of little concern to them will affect their share of the vote (one former Leader was not even aware that his district had a buildings-at-risk register, and was even less aware of the powers that could be taken); secondly, from officers without experience of using the powers and afraid of lumbering their authority with a hefty repair bill for a building; and thirdly, from the councils' legal teams, who are usually already overloaded and not necessarily skilled in dealing with these powers.

However, the benefits to the historic environment of the use of these powers are obvious, and there is also serious political capital to be made from their effective use. SAVE has experienced the use of these powers itself with Castle House in Bridgwater (though fortunately not from the wrong side). Castle House is a Grade II* listed building, dating from the 1850s. Its owner bought it blind at auction for around £2,000, believing it was a bargain, possibly a site for a new building. The building was left to rot, and were it not for the tenacious actions of the planning officer, Mark Alcock, at Sedgemoor District Council, it probably would have collapsed: indeed one councillor had offered to assist its early demise.

Figure 8 Castle House in Bridgwater is SAVE's latest challenge – an 1850 experimental concrete building in need of complete restoration, having been subject to both Urgent Works and Repairs notices under its previous owner.

Urgent Works notices were served and a scaffold erected to protect the building from the elements. The owner moved the building from company to company to avoid liability for the works, and then SAVE's architect, Kit Routledge of Richard Pedlar Architects, drew up the specification for a Repairs Notice. This was duly served, but the owner again moved companies, before finally being advised by his lawyers that any further action was futile, and would have resulted in a Compulsory Purchase Order (CPO). The building was then passed to SAVE, with both planners and politicians being pleased for the building to be in the hands of an organization committed to its repair. The members and legal team are now familiar with the process and hopefully confident enough to be able to use it in the future to help solve the problems surrounding other individual buildings-at-risk, and SAVE has a new restoration challenge.

One way of ensuring that local authorities take action on buildings-at-risk would be to implement the use of buildings-at-risk registers, with the consequent level of action on the local authority's part being made into a Best Value Performance Indicator. This would also help national groups (including building preservation trusts) understand where their efforts might be best focused. Every local authority should be encouraged to have a buildings-at-risk register as a simple matter of best practice: in 2005 only around 30% of authorities held such information.

Maintenance

SAVE holds that the most sensible form of preservation of the historic fabric of the nation is regular maintenance. However, at present there is little to encourage this. Indeed it could be argued that the current grant regime is perverse in that it rewards those who allow their buildings to fall into serious disrepair. SAVE holds that there remains a fundamental need for a paradigm shift in the focus of the conservation world from repair to maintenance.

This has been noted by the Heritage Minister, David Lammy, who at Heritage Link's 2005 open day used the phrase 'invest to save'.[17] While this recognition of the problem is a positive step, action would be even more welcome. To encourage maintenance, there needs to be some form of financial incentive: at present it is penalized through the punitive VAT regime.

There are a number of pilot maintenance schemes under way. These are focused on places of worship; partly on the basis of need, and partly because through the quinquennial inspection scheme there is a framework on which to build. The basis for these, in St Edmondsbury Diocese, London Diocese and Gloucester Diocese, is an annual maintenance inspection, with minor works carried out and recommendations made to those responsible for the building. These schemes fit together with English Heritage's recent launch of its 'Inspired!' campaign, which raised the prospect of small grants for church maintenance – a move that is to be welcomed.

A wide-ranging pilot scheme was carried out in Bath in 2002–3 by Maintain our Heritage (MoH), testing the practicalities, from access to health and safety, of a maintenance inspection scheme on a range of different buildings. It showed that while there is demand for such a service, the initial costs are high. Following this, MoH carried out a large research project, sponsored by the Department for Trade and Industry (DTI),

Figure 9 Testing out different methods of access during Maintain our Heritage's Bath pilot maintenance scheme.

Figure 10 The Welsh Streets in Liverpool contain a community divided by the demolition proposals foisted upon them into those who want new houses and those who want to stay in their houses. A better thought out scheme could quite easily reconcile situations like this.

English Heritage, the Heritage Lottery Fund and others covering best practice, owner attitudes, the provision of commercial maintenance services, prevention, inspection and maintenance technologies, the business case, and maintenance education.[18]

Pathfinder / Housing Market Renewal Initiative

The greatest threat to our heritage (other than the effects of wind and rain) is Ruth Kelly's Department for Communities and Local Government, in the form of its Housing Market Renewal Initiative (HMRI). This policy will, at current rates of demolition, see the destruction of over 57,000 homes, the majority of them pre-1919, and their attendant communities.[19] These buildings represent an accessible and ownable heritage, albeit one completely undervalued and unappreciated by those in control of HMRI.

The scheme's basic intentions are good – to raise the quality of housing in these areas, to bolster the housing market where it is weak and to tackle the social problems the areas face. However, rather than proposing imaginative ways of tackling these problems, the blunt tool of demolition is being utilized, and the communities' viewpoints are not being taken into account – many of the consultation exercises so far have been more a case of expensive consultants telling locals what to expect. This has understandably created a mass of opposition and wave of revulsion amongst sections of the national press. SAVE has led the fight at the national level, publishing its 2006 report, *Pathfinder*, which looked at the problems and the possible solutions, as well as opportunities to stitch back the townscape.

One of the smaller scale demolition proposals – for 160 houses (others are for up to 1,500) – in Darwen, Lancashire, illustrates much of what is wrong with the policy. In a consultation, residents of this area of 1860s stone-built terraces were asked if they would like their area to be 'regenerated' or for nothing to happen. Naturally most opted for regeneration, without it being explained that this would involve them moving out of their homes and the area being flattened. Of the 160 houses, at the start of the scheme under 10% were vacant (most of which were owned by the local authority). External condition surveys were carried out, taking no more than ten minutes per house. On the basis of these, the houses were condemned as structurally unsound.[20] Structural engineer Professor Brian Clancy was called in by 16 residents who were looking for a second opinion. Professor Clancy, a chartered structural and civil engineer (and a past president of the Institution of Structural Engineers and past chairman of the Association of Consulting Engineers), examined 15 of the condemned houses and could find nothing particularly wrong with them: houses were being condemned on the basis of slipped tiles and so forth.

Figure 11 Standing outside her 1860s stone-built terrace in Darwen, this elderly woman and her husband faced eviction from their own home as the market in the area was deemed to have failed, in spite of evidence to the contrary. Theirs is one of 160 houses facing demolition to make way for a city academy in the Red Earth Triangle.

As soon as it became clear that the area was potentially up for demolition, blight set in. Social landlords started to move tenants out in anticipation, and bricked up the windows of their properties. Basic council services to the area declined, and those opposing demolition were, in effect, obstructed by the local authority in their quest to discover what was going on – for example it refused to produce figures illustrating the cost of restoration versus demolition. Then the residents received letters announcing the compulsory purchase of their houses – including people who had lived there for over 60 years. House prices in the area before the announcement of the scheme had been buoyant by local standards. The CPO went to a public inquiry, where the locals and SAVE put up stiff opposition to the clearance proposals. The Inspector over-ruled the CPO.

Conclusion

In spite of the more subtle new threats to the historic built environment, a review of the SAVE publications back catalogue (with over 120 entries) shows that many issues remain the same, and depressingly, that many of the lessons have still not been learnt. The historic built environment in the

UK is far from safe and secure – familiarity continues, in the minds of some, to breed contempt. Heritage is a mere footnote for government, and were it not for the energy and enthusiasm of the many groups in the sector among which SAVE is numbered, it would be even less.

There will always be a role for SAVE, whether acting to prevent the demolition or mutilation of historic buildings and places, thinking up new ways of reusing them, or in fostering interest in different ways of encouraging preservation; whether in heritage-proofing public policy, placing a duty of care on local and national government in relation to its building stock, looking at Swiss-style mortgages (taken out over 100 years, thereby encouraging proper maintenance), understanding how to undo the damage of the past, or considering tax breaks for those bringing redundant historic buildings back into use. These are but a few of the areas SAVE might consider using its open remit to place on the agenda in the future.

Biography

Adam Wilkinson MA, MSc
Adam Wilkinson has been Secretary of SAVE Britain's Heritage since 2001, having previously worked at UNESCO in Paris. He serves on the Churches Conservation Trust Conservation Advisory Committee, is a Trustee of the Mausolea and Monuments Trust, a director of Maintain our Heritage and the author of several SAVE reports.

Notes

1 Binney, M. and Watson Smyth, M., 'The Origins of SAVE' in *The SAVE Britain's Heritage Action Guide*, Collins and Brown, London (1991).

2 SAVE Trustees and Advisory Committee by occupation as of August 2006 (including the SAVE Trust): 1 architect, 3 journalists, 2 planners, 1 conservation officer, 1 private practice, 5 architectural historians, 2 curators, 1 LA councillor, 3 developers, 1 financier, 2 other.

3 Compare Powell, K., *What! Conservation in Gateshead*, SAVE, London (1982); with Palmer, E., *Damned Beautiful*, SAVE, London (2005).

4 The most complete history of the organization is: Binney, M., *SAVE Britain's Heritage 1975–2005: 30 Years of Campaigning*, Scala, London (2005). Further information on SAVE's campaigns can be found at www.savebritainsheritage.org (accessed 15 September 2006).

5 'Protecting and Preserving our Heritage', House of Commons Culture, Media and Sport Committee 3rd Report of the Session 2005–6, Vols I–III, July 2006.

6 The 'Clews' report, unpublished, n.d.

7 Wilkinson, A., *Enough has been Bulldozed! SAVE Farnborough, the Cradle of British Aviation*, SAVE, London (2002).

8 DCMS press release 79/06.

9 For an excellent discussion of the failings of PFI in relation to publicly owned historic buildings, see Pollard, R., *Silence in Court – The Future of the UK's Historic Law Courts*, SAVE, London (2004).

10 'Protecting and Preserving our Heritage', House of Commons Culture, Media and Sport Committee 3rd Report of the Session 2005–6, Vols I–III, July 2006.

11 Ibid.

12 In an unfortunate reflection on the state of affairs, the organizations that have given SAVE examples such as these have been asked that details of the projects are withheld.

13 National Conservation Salaries Monitoring Database (1998 onwards) for IHBC (unpublished).

14 SAVE Britain's Heritage Newsletter, Nov 2004.

15 In response to a Parliamentary Question (HL1205) in March 2001, Lord Whitty promised changes to the GPDO shortly. This promise has been repeated regularly since by other junior ministers, in spite of a complete lack of action.

16 For more information on the use of S215 notices, see: Town and Country Planning Act 1990, Section 215 Best Practice Guidance, http://www.communities.gov.uk/pub/468/TownandCountryPlanningAct1990Section215BestPracticeGuidancePDF 508Kb_id1144468.pdf (accessed 15 September 2006).

17 Minute of 2006 Heritage Link AGM. Yet to be published.

18 This research and a report on the Bath pilot can be downloaded from www.maintainourheritage.co.uk/pilot.htm (accessed 25 September 2006).

19 Para 9.19 of 'Moving Forward the Northern Way', Sept 2004, www.thenorthernway.co.uk/report_sept04.html (accessed 15 September 2006).

20 Full evidence on this was provided to the public inquiry by Natasha Lea Jones on behalf of the local amenity group 'Heart', Brian Clancy, individual local residents and SAVE Britain's Heritage.

Scientific Research into Architectural Conservation

Peter Brimblecombe and Carlota M. Grossi

Abstract

A comprehensive research agenda is needed that balances past achievements with emerging issues of the future. Although the strengths of current research are worth highlighting, it is equally important to understand the weaknesses and major gaps in the research on conserving the built environment. In particular, the lack of an overall research strategy, inadequate funding and poor translation of research into policy and practice are key problems. Strategies for the future need to allow the research agenda for the historic environment to be both focused and effective. An analysis of European research can be found in the EU Parliament's Scientific and Technological Options Assessment Panel *(STOA) report of 2001:* Technological requirements for solutions in the conservation and protection of historic monuments and archaeological remains,[1] *while in the UK, English Heritage has published its strategy as* Discovering the Past, Shaping the Future.[2] *There is a range of new research needed, from work on philosophical, social and management issues to science and technology. The field also continues to lack good publications for communicating its research output to the end users.*

Introduction

Research in Europe, and more particularly the UK, has been transformed by changes in recent years. The Lisbon agenda (2000) for European growth and competitiveness affirmed the desire for Europe to become a leader in science, technology and education. It also opened the possibility of more fundamental research within the European Research Area (ERA) and suggested a structure for the new European Commission Framework Program 7. In the UK the Research Assessment Exercise (RAE) has focused academic institutions on clear output measures for their research,[3] most notably the production of high quality papers in leading journals. The Treasury report *Science and Innovation Investment Framework 2004–2014*[4] has looked beyond the RAE with an interest in a simpler approach to evaluating university research using various metrics (research income, publications, etc.). In the area of conservation research, English Heritage has published a strategy document *Discovering the Past, Shaping the Future* in 2005,[5] which addresses not only a vision for their own research ambitions, but also more widely the question of heritage research in the UK. In the spring of 2006, the House of Lords' Science and Technology Sub-Committee on Science and Heritage heard evidence of the state of heritage research in the UK; and English Heritage and the Research Councils promoted a meeting in Birmingham entitled 'Preserving our Past' with the goal of defining clusters that would investigate inter disciplinary research areas concerning the historic environment.

European heritage research

In the last decades of the twentieth century, the European Union supported research on cultural heritage through a range of initiatives, perhaps most notably the Framework Programmes. These encouraged the development of pan-European groups of researchers and important new areas of collaborative research. Initially, there was a focus on the built heritage, but this research effort has become more diverse over time. There have been numerous high-quality publications, spin-offs into other research fields and imaginative efforts at dissemination.[6] Nevertheless, there has been criticism of the variability and quality of the output of such European research and its translation into policy. However, the Department of Trade and Industry (DTI) study, *The Impact of the EU Framework Programmes in the UK* (2004),[7] could offer no viable alternatives to the methodologies of the European programmes. From the heritage perspective, there is a feeling that the earlier Framework Programmes placed European heritage research in a pre-eminent position, as outlined in the STOA report, but that this support has declined over time. Thus, the arrival of more recent programmes, notably FP6, shows declining funding for heritage research.

In the emerging Seventh Framework Programme (FP7), there are hopes that cultural heritage tasks will be identified clearly within the key actions, perhaps under the heading of Environment and Climate Change. The creation of a European Research Area (ERA) is linked to a strong desire for a European Research Council (ERC) to support frontier research. However, heritage research is not often viewed as offering fundamental insights, so it may not be well funded by any ERC that emerges. There will, however, be continued support for training through the Marie Curie scheme, and an interest in fostering industrial research by encouraging the involvement of Small to Medium Enterprises (SMEs).

We have to recognize that research funded by the European Commission is essentially collaborative, and often needs to demonstrate support for EU policy and to be pan-European in nature. This clearly means that both the development of more speculative theoretical ideas and any focus on collections of specific national interest tend to be neglected. National programmes have emerged in some European countries, such as Italy. In others, such as the UK, national funding is limited and fails to promote a critical mass of research. Funding here relies on limited programmes within the research councils, and on support from charities such as the Leverhulme Trust.

Current situation in the UK

The need for strategic plans for heritage research has been recognized, and English Heritage has been active in laying out a research strategy in recent years (as in their 2005 publication *Discovering the Past, Shaping the Future*[8]). More recently, the House of Lords Science and Technology Sub-Committee on Science and Heritage has heard evidence about the heritage research sector. This recent development of heritage research strategies within the UK ought to lead to stronger research funding. The strategies recognize the importance of past achievements and the large and productive contribution to European programmes by UK researchers. Success is seen in the substantial flow of academic publications from research groups and their involvement in major conferences reviewing European heritage research, e.g. the conference on *Sustaining Europe's Cultural Heritage* held in London in September 2004.

Research in the UK takes place in a range of academic and research institutions, governmental and related UK organizations and consultancy groups, and is carried out by a variety of practitioners. Obviously the more theoretical work is typically done in the university environment, often in collaboration with conservators, while research in heritage organizations is often directed towards more specific and practical problems. This is a natural and sensible balance. Although the overall UK research output is often of very high quality, it lacks a sense of overall direction and mechanisms for converting its findings into practice. Organizations such as ICON (Institute of Conservation) or specialist professional groupings such as SWAPNET (Stone Weathering and Atmospheric Pollution Network) are keen to see research output used, but this has remained a difficult process.

There is a certain cynicism in some quarters that research remains a mere academic exercise, detached from reality, and serving only to promote further research. However, the Environmental Research Funders' Forum, which brought together the main UK public bodies which fund or use environmental research, gained much insight into what might be potentially useful research by engaging with research end users.[9] This enabled them to identify a range of practical issues that are also relevant to a heritage research strategy. It is clear that the conflict between policy-driven research and academically driven research needs to be addressed. Often, researchers failed to appreciate the time pressures involved in policy-making, but conversely some researchers felt under pressure to deliver results too quickly. Research councils often found it difficult to meet the needs of end users. One suggestion was to increase policy makers' involvement in research, but this can demand large time commitments from them.

Research gaps

Any planning for future research policy has to recognize not only the strengths of past approaches but also the weaknesses of the UK's strategy. In the past, it has not been clear how the research agenda was set, or how the quality and advancement of knowledge was assessed in heritage research.

Recently English Heritage has laid out seven equally important themes for research on the historic environment:[10]

- discovering, studying and defining historic assets and their significance;
- studying and establishing the socio-economic and other values and needs of the historic environment and those concerned with it;
- engaging and developing diverse audiences;
- studying and assessing the risks to historic assets and devising responses;
- studying historic assets and improving their preservation and interpretation;
- studying and developing information management;
- studying and devising ways to make English heritage and the sector more effective.

These themes are meant to develop into research programmes, but will naturally need refining in terms of the specification of more tactical objectives rather than such broad aspirations.

The European Commission's collection of 'Expressions of Interest' (these are inputs invited from researchers by the Commission to help prepare Framework Programmes) reveal their desire to see more research on sustainability and climate rather than pollution impacts, carried out by small active research groups rather than large networks.

Philosophical notions

There is a range of important intellectual questions around heritage research concerned with philosophical notions, such as: aesthetics, authenticity, value and inclusiveness. Access is a key issue in many areas of socio-economic planning, but for heritage conservation this creates both benefits and problems. Increasing access may also impose on us a greater acceptance that heritage cannot last forever, along with the need to devise new ways of valuing culture and its use and new methodologies for assessing risk. In a broader sense, these issues also shift the focus away from the individual monument towards its cultural environment. Increasing visitor flow increases the need to understand tourism, so there has be a rising interest in eco-tourism or sustainable tourism.

There is also increased interest in the very nature of conservation, addressing questions concerning the value of preservation, and the notion of preventive conservation, restoration and reversibility. Although these are important research topics, it will be necessary to ensure that they move beyond scholarly discourse and lead to outcomes of relevance to conservators and heritage managers. The best heritage research usually involves close co-operation with practitioners to address key conservation issues. In addition, there need to be discussions about the integration of scientific research with evolving policy. This has meant that increasingly, research projects run workshops on completion.

Conservation needs new tools to assist in practical matters. Important ones, such as: indicators of current condition, predictors of future condition, optimum maintenance, repair frequency and compatibility, are inherent in a range of current evaluations that should feed into future directions for sector research (e.g. the European Commission INCO project Prodomea ICA3-CT-2002-10021a).

Management guidance

There is also a range of guiding principles relevant to management of heritage that needs more research. This would include an understanding of the precautionary principle, sustainability and life-cycle analysis. These ideas are widely known, but not always tailored to make them useful in terms of our built heritage. More research is required into developing standards for the protection of this heritage. This would include standards for allowable exposure, repair materials, maintenance and restoration. This research also has to recognize a shift in focus from individual objects to large-scale entities, such as the landscapes that surround heritage 'objects'. Future developments of building codes, restoration codes, listing revisions, training requirements and training courses are going to need the outputs from such research.

Some approaches to sustainability in the heritage sector are applied in narrow and restrictive ways. It is often solely in terms of energy that sustainability has been seen in very quantitative approaches. For example, Graedel and Klee have written of the importance of quantifying our approaches to environmental sustainability if they are to have a real chance of success.[11] However, if handled too prescriptively, these may avoid more general sociological understandings and end up expressing everything in terms of kilowatt-hours. This can limit policy development, because it avoids the broader context. In terms of the built heritage, the idea of saving energy, especially energy related to heating and lighting gains in new buildings, has also to take into account factors such as the embedded energy in older buildings, and the benefits of local traditional materials.

The concept of 'life-cycle analysis' (an assessment of impact based on an analysis that considers the real costs of each stage of a product's life from extraction of materials, through creation to disposal) has tended to follow the same quantitative pattern, and has been most concerned, as one might imagine, with energy issues. The 'precautionary principle' is a further tool related to the concept of sustainability, but should be expressed in terms of cultural heritage. In environmental issues this principle requires that where there is reasonable suspicion of harm, lack of scientific certainty or consensus, it should not postpone preventative action. This does not completely cover the concerns within heritage conservation, where there is often a need to minimize intervention and invasive cleaning. Additionally there have been advances in environmental economics that have allowed non-monetary costs to be accounted for in many areas, but the economics of heritage conservation requires a specific understanding based on its unique status to give insight into improving its management.[12]

Scientific research

There is a pressing need to assess the magnitude of future threats to the historic environment. The importance of climate change is now recognized, but threats are much broader than this, and include social and political factors associated with our changing world. Terrorism is one threat[13] that has received special prominence in recent years. There are increasing threats to monuments and those who visit them,[14] but the subject is rarely considered in academic study.

On a more technical level, more research is required to understand the environment–materials interface, and areas such as buried heritage is too easily neglected, e.g. the effect of climate change on the Roman ruins at Alchester.[15] The importance of understanding new materials, new building types and new techniques has to be recognized, while remembering the importance of traditional materials and techniques. There are new scientific and analytical methods, such as the increasing use of synchrotron radiation to explore the structure of organic materials at a molecular level. Some new techniques offer the possibility of non-destructive and unobtrusive approaches to monitoring both the state of the environment and our heritage.

Dissemination issues

Research is of little value unless it is properly disseminated and used. There are a number of problems in communicating the outcomes of heritage research and getting it translated into policy.

At the academic level, there is a paucity of suitable high-impact journals in the heritage research area, because their citation and publishing traditions fall outside the sciences. Some newer databases are broadening their range of content, however, and this may change. Problems remain: although two popular journals in the area, *Journal of Cultural Heritage* and *Studies in Conservation*, are on databases such as Web of Knowledge and *SCOPUS*, they remain poorly cited (although in mid 2006 *Journal of Cultural Heritage* scored an impact factor of 1.7). *The Journal of Architectural Conservation* is not included on either database. Academic researchers wish their articles to be widely read and cited, but the absence of conservation-related journals on important databases hinders widespread awareness of work published in these. It also lowers the incentive for university researchers to publish in the conservation literature. They will often choose to publish in the mainstream geological, environmental and chemical literature, which can give a higher profile to their work, but this often means that it fails to reach conservators. Much of the conservation literature that influences practitioners derives from conference proceedings. Although fine articles appear here, the refereeing is often cursory, and the quality of the output variable.

The lack of a common language between scientists, conservators and managers has also been seen as inhibiting the conversion of research outcomes into useful policy. It is reflected in the continuing problem of getting well-trained scientists to cross into the conservation field. While managers of heritage realize the importance of research, the day-to-day pressures of their job often make it hard for them to make the time to take on relevant research output in order to change practice.

Conclusions and future

Europe has developed an active research community producing results of the highest international quality in heritage research. These are revealed in excellent academic publications, but there are few widely read journals that bring research output to the attention of conservators and heritage managers.

There have been some imaginative attempts at dissemination in conferences and courses, but more effort is needed here. Research has had important spin-offs and made contributions to emerging European legislation, but its impact needs to increase in the future. There is a growing desire for the widest stakeholder involvement in the decision-making process; for example, the Environmental Research Funders' Forum noted on the issue of stakeholder views of the environment: 'Politically, there is a problem that environmental science is not seen as being of great economic value. Its true economic worth is not being appreciated.'[16] This statement will have much resonance with those concerned about research into the built heritage.

It is hoped that recognition of cultural heritage within Framework Programme 7 will ensure a continued development of a European heritage research community. There is a desire to develop a coherent community and focus on relevant science. However, pan-European support should be at an appropriate scale and flexibility (i.e. 1 million Euro rather than 20 million Euro projects). Current discussions in the UK reveal the need for national support of heritage research.

The English Heritage research agenda places much emphasis on the built heritage. The agenda is not narrowly confined to the scientific analysis of materials, processes or threats, but lays out explicit objectives for understanding the social context of our heritage, information management and good practice within the sector. These are broad and noble ambitions that now demand clear tactics if the strategic objectives are to be met.

Biography

Professor Peter Brimblecombe BSc, MSc, PhD
Peter Brimblecombe is a professor of atmospheric chemistry at the University of East Anglia with a particular interest in the relationship between air pollution and cultural heritage.

Dr Carlota M. Grossi BSc, MSc, PhD
Carlota Grossi is a senior research associate at the University of East Anglia. Her speciality is building stone decay and conservation.

Notes

1 *Technological requirements for solutions in the conservation and protection of historic monuments and archaeological remains* Report (STOA Project 2000/13-CULT/04), STOA (2001), www.ucl.ac.uk/sustainableheritage/STOA_report.pdf (accessed September 2006).
2 *Discovering the Past, Shaping the Future. Research Strategy 2005–2010*, English Heritage (2005).
3 RAF, *UK Research Assessment Exercise* (2006), www.rae.ac.uk/default.asp (accessed August 2006).
4 HM Treasury, Department of Trade and Industry and Department for Education and Skills, *Science and Innovation Investment Framework 2004–2014* (2004), www.hm-treasury.gov.uk/spending_review/spend_sr04/associated_documents/spending_sr04_science.cfm (accessed April 2006).
5 English Heritage, *op. cit.* (2005). [Note 2]
6 English Heritage, *op. cit.* (2005). [Note 2]
7 *The Impact of the EU Framework Programmes in the UK*, Consultation document on the 7th EU R&D Framework Programme, Department of Trade and Industry (2004), www.dti.gov.uk/files/file19063.pdf (accessed September 2006).
8 English Heritage, *op. cit.* (2005). [Note 2]
9 *Environmental Research Funders' Forum*, ERFF (2003), www.erff.org.uk/documents/Finalversion.pdf (accessed April, 2006).
10 English Heritage, *op. cit.* (2005). [Note 2]
11 Graedel, T. E. and Klee, R. J., 'Getting serious about sustainability', *Environmental Science and Technology*, 36 (2002), pp. 523–529.
12 Lithgow, K., Lloyd, H., Brimblecombe, P., Yoon, Y. H. and Thickett, D., 'Managing dust in historic houses – the visitor/conservator interface', *ICOM Committee for Conservation, 14th Triannual Meeting*, Volume II (2005), pp. 662–669.
13 Coningham, R. and Lewer, N., 'Paradise lost: The bombing of the Temple of the Tooth – a UNESCO World Heritage site in Sri Lanka', *Antiquity*, Vol. 73 (1999), pp. 857–866.
14 STOA, *op. cit.* (2001). [Note 1]
15 Sauer, E., 'Alchester – in search of Vespasian', *Current Archaeology*, 196 (2005), pp. 168–176.
16 ERFF, *op. cit.* (2003). [Note 9]

Appendix

The Listing of Buildings

Bob Kindred

This Appendix gives a brief general outline of the way in which 'listed' buildings in England are protected.[1] Several commentators have dealt with its evolution at length.[2]

Introduction

It was during World War Two that the government first legislated for a 'statutory list' of buildings of special architectural and historic interest, and the process commenced in 1946–7.[3] The initial compilation took the government appointed investigators well over twenty years to complete, mainly because of a lack of ministerial commitment and there were often fierce criticisms of what was protected and what was not.[4]

The initial programme which ran until 1968, produced lists that were undoubtedly patchy – especially the earlier ones – and a second nationwide resurvey was therefore undertaken starting in 1969. The process was accelerated following the notorious demolition of the Firestone Factory[5] in west London in 1980 and the remaining fieldwork done on an area-by-area basis was completed by the spring of 1989. Responsibility for compiling the lists was transferred outside government with the establishment of English Heritage in 1984.

Subsequently the focus switched to more precisely targeted studies of building types which analysis had indicated were under-represented in the lists.[6] In recent years there has been an emphasis on identifying key buildings of the twentieth century for protection, including those of the post-war period; and at each stage recommendations for protection of specific buildings came forward.

Selection criteria

The basis for determining what should be protected (the selection criteria) has evolved considerably since it was first introduced in 1946 and the most recent government guidance dates from 1994[7] (although this is currently under further review).

- *architectural interest:* the lists are meant to include all buildings which are of importance to the nation for the interest of their architectural design, decoration and craftsmanship; also important examples of particular building types and techniques (e.g. buildings displaying technological innovation or virtuosity) and significant plan forms;
- *historic interest:* this includes buildings which illustrate important aspects of the nation's social, economic, cultural or military history;
- *close historical association:* with nationally important people or events;
- *group value:* especially where buildings comprise an important architectural or historic unity or a fine example of planning (e.g. squares, terraces or model villages).

Not all these criteria will be relevant to every case, but a particular building may qualify for listing under more than one of them.

Application of the criteria

As a consequence of many years of application, PPG 15 makes clear that the older a building is, and the fewer the surviving examples of its kind, the more likely it is to have historic importance and that the following buildings should now be listed:

- *before 1700:* all buildings which survive in anything like their original condition;
- *1700 to 1840:* most buildings though some selection is necessary;
- *1840 to 1914:* only buildings of definite quality because of the greatly increased numbers surviving, greater selection is needed to identify the best examples of particular building types;
- *after 1914:* only selected buildings;
- *between 30 years and 10 years old:* only buildings which are of outstanding quality;
- *less than ten years old:* not listed.

This process has resulted in approximately 500,000 buildings (and other above ground structures) being listed – representing about 2% of all buildings in England.

Grades

Not all buildings are of equal worth. Listed buildings are generally classified into three grades: Grade I, Grade II* and Grade II although this aspect is non-statutory.

- Grade I buildings (representing about 2.5% of the total) are those of exceptional interest;
- Grade II* buildings (representing about 5.6%) are particularly important buildings of more than special interest;
- The remainder (representing 91.9%) are Grade II buildings of special interest, warranting every effort to preserve them.

There are claims that this system, which has evolved over 60 years, is not comprehensible to the public and that a new one with revised and elaborated criteria, and revised and simplified grades is required. Government and English Heritage has now spent several years evaluating and consulting on potential changes, but some of these will require legislation to be implemented and may take many further years to fully introduce, while the existing and new arrangements operate in parallel. For the present the inventorization of historic buildings in England will continue in the form outlined above.

Notes

1   The same general principles apply in Wales and Scotland.
2   Readers are referred to Earl, J., *Building Conservation Philosophy*, third edition, Donhead, Shaftesbury (2003), particularly Appendix 6 on the initial instructions to investigators; Delafons, J., *Politics and Preservation – A Policy History of the Built Heritage 1882–1996*, E & FN Spon, London (1997); Acworth, A. and Sir Anthony Wagner, '25 Years of Listing', *Architectural Review*, Nov. 1970; Harvey, J. *et al*, 'The Origin of Listed Buildings' *Ancient Monuments Society Transactions*, Vol. 37 (1993).
3   Local authorities had been given powers in planning legislation in 1932 to prepare local lists of buildings of special interest but very few did so.
4   Earl, ibid, p. 195.
5   This fine Art Deco building of 1928 was demolished over an August public holiday by the owner/developer Trafalgar House, the day before it was due to be protected. The relevant government minister, Michael Heseltine, accelerated the protection process with impressive results.
6   For example Lancashire Cotton Mills and Yorkshire Wool Mills.
7   Department of the Environment/Department of National Heritage, *Planning Policy Guidance Note 15, Planning and the Historic Environment (PPG15)*, HMSO, London (1994) paragraphs 6.10–6.16.

JOURNAL OF
Architectural Conservation
The international journal for historic buildings, monuments and places

Patron: Sir Bernard Feilden

Consultant Editors:
Professor Vincent Shacklock
Elizabeth Hirst
Professor Norman R. Weiss
Bob Kindred MBE

The scope of this international journal is intended to be wide-ranging and include discussion on aesthetics and philosophies; historical influences; project evaluation and control; repair techniques; materials; reuse of buildings; legal issues; inspection, recording and monitoring; management and interpretation; and historic parks and gardens.

© Donhead Publishing 2006

Editorial, Publishing and Subscriptions
Donhead Publishing Ltd
Lower Coombe, Donhead St Mary
Shaftesbury, Dorset SP7 9LY, UK
Tel: +44 (0)1747 828422
Fax: +44 (0)1747 828522
E-mail: jac@donhead.com
www.donhead.com

Managing Editor: Jill Pearce
Publishing Manager: Dorothy Newberry

Journal of Architectural Conservation is published three times per year.
Annual institutional subscription 2007: £96.00
Annual personal subscription 2007: £47.00
Back volumes and single issues are available from Donhead Publishing.

Papers appearing in the *Journal of Architectural Conservation* are indexed or abstracted in *Art and Archaeology Technical Abstracts*, *Avery Index to Architectural Periodicals*, *British Humanities Index*, *Art Index* and *Getty Conservation Institute Project Bibliographies*. Abstracts to all papers can be viewed on the Donhead website: **www.donhead.com**

ISSN: 1355-6207

Printed and bound by CPI Group (UK) Ltd, Croydon, CR0 4YY

23/10/2024

01777680-0004